Nuclear Power - Control, Reliability and Human Factors

Nuclear Power - Control, Reliability and Human Factors

Editor

Azat Solak

scitus
academics

Nuclear Power - Control, Reliability and Human Factors
Edited by **Azat Solak**

ISBN: 978-1-68117-513-3
Library of Congress Control Number: 2016942873

SCITUS Academics LLC,
www.scitusacademics.com
Box No. 4766, 616 Corporate Way,
Suite 2, Valley Cottage,
NY 10989

Printed in United States of America on Acid Free Paper

Preface

Advances and reliability in reactor designs, materials and edges of human machinery ensure the safety and reliability of evolving reactor technology and reduce the possibilities for serious human errors such as those of the past. Nuclear power can be the use of nuclear reactions to generate heat from nuclear energy, which is most often used in steam turbines to generate electricity in a nuclear power plant. The term includes nuclear fission and nuclear fusion. A nuclear reactor, formerly known as a nuclear pile, is a device used to initiate and control a sustained reaction of the nuclear chain. Nuclear reactors are used in nuclear power plants to generate electricity and to propel vessels. New control and instrumentation technologies based on digital systems, new sensors and measuring methods enable nuclear options to be cost-effective, reliable and safe. Autonomous operating circumstances are gradually becoming prevalent for small modular systems. Nuclear technology uses the energy generated by the separation of the atoms of specific elements. It was first produced in the 1940s and during the Second World War until 1945 research focused initially on the production of bombs by dividing the atoms of certain isotopes of either uranium or possibly plutonium. In the 1950s, attention was focused on the peaceful purposes of nuclear fission, especially the generation of electricity. Today, in the first years of nuclear power, the world produces as much electricity from nuclear power as it did from all sources. Civil nuclear power now has more than 16,500 reactor years of experience and supplies nearly 11.5% of the world's electricity requirements from reactors in thirty-one countries. In reality, much more than those countries depend on nuclear power through regional grids. Many countries have also built research reactors that provide a source of neutron beams for industrial and medical isotopes and for scientific research. The nuclear power plant lies at the border between the greatest expectations of humanity and its deepest fears for the future. Atomic energy comes in one hand with a thoroughly clean energy alternative that safeguards us from the fossil fuel dependency. On the other hand, it summons disaster images.

This book provides details on key topics related to instrumentation and control, operational reliability, system aging and interfaces between people and machines. The book should be of enormous value to a broad group of prospective readers –academics, researchers and industry specialists– interested in learning about nuclear power.

Table of Contents

Chapter 1

Integrated Approach for Actual Safety Analysis

Francesco D'Auria[1], Walter Giannotti[1] and Marco Cherubini[1]

[1] GRNSPG - University of Pisa,, Italy

1. INTRODUCTION

Actual trend in reactor safety deterministic analysis are evolving toward best estimate approach. Best estimate analyses imply use of best estimate codes and input data. The best estimate concept is not limited to thermal-hydraulics rather in general terms it covers many fields, likewise three dimensional neutron kinetics, structural analysis and containment performance evaluation.

The general frame is to put efforts in avoiding conservative assumptions performing analysis adopting the best tool available for each specific topic, all contributing to give an integrated evaluation of the plant response.

The needs to adopt an integrated approach in performing safety analysis come from the inherent complexity of a Nuclear Power Plant and from the tight interactions among the subsystems constituting the plant itself. These interactions directly involve the necessity to consider a broad spectrum of disciplines typically coming into play in different not interacting analyses.

An example of the integral approach is given in the present document. The integral approach has been pursued for the safety analyses of the 'post-Chernobyl modernized' Reactor Bolshoy Moshchnosty Kipyashiy (RBMK) specifically for Smolensk 3. These analyses were performed at the University of Pisa within the framework of a European Commission sponsored activity.

The mentioned analyses deal with events occurring in the primary circuit, as well as excluding those events originated from plant status different from the nominal operating conditions. Following the evaluation of the current state of the art in the safety analysis area, targets for the analysis were established together with suitable chains of computational tools. The availability of computational tools, including codes, nodalisations and boundary and initial conditions for the Smolensk 3 Nuclear Power Plant, brought to their application to the prediction of the selected transient evolutions that, however, are not classified as licensing studies.

The integrated approach for safety analysis yields to the evaluation of complex scenarios not predictable adopting just a single computational tool. Example is given considering the Multiple Pressure Tube Rupture (MPTR) event which constitute one of the main concern of this kind of plant.

The content of this document includes an introduction to the critical issues to be accounted for in the frame of an integral safety analysis approach; the selection of suitable computational tools to proper deal with the scenario subject of the investigation; an approach on how to link (coupling issues) the selected tools; the use of intermediate code outcomes and interpretation of the global predicted plant behaviour. All the aspects presented in general terms are applied in the case study of a Multiple Pressure Tube Rupture having as reference plant the Smolensk 3 Nuclear Power Plant. The selected event may occur as a consequence of a fuel channel blockage which (if not detected) brought to the rupture of the affected pressure tube. The dynamic loads generated by its breach may lead to the rupture of the surrounding pressure tubes. Direct consequence of the pressure tube rupture is the pressurization of the reactor cavity which envelopes all the core. In the case of Multiple Pressure Tube Rupture event, involving a large number of pressure tubes, the lifting of the reactor cavity top may occur, putting in direct connection the core with the environment. The present example is a kind of analysis that cannot be performed if an integrated approach is not adopted.

2. FRAMEWORK

The best estimate approach is the actual trend of the NPP deterministic analysis (International Atomic Energy Agency [IAEA], 2008). The concept of best estimate is generally applied to the software codes used in the analysis. However the best estimate approach concept has a broader meaning. It applies to the general framework of the analysis, and it involves not only the codes, but the kind of analyses to be performed, the approach to realize the models to be realized for the analyses, the input data including boundary and initial conditions also. The best estimate approach is not only connected with a calculation performed with a best estimate code. The result of the analysis is a best estimate evaluation, if all the aspects of the analysis (input data, systems models, results) are best estimate, in addition to the codes. As a consequence the use of a best estimate code, assuming not best estimate data or systems model cannot be considered a best estimate analysis.

A calculation of a complex system like a NPP, poses a lot of issues to perform a best estimate analysis. The main relevant aspect is constituted by the many areas involved in the analysis of a NPP. Knowledge in many technical areas are necessary. The solution can be obtained by "linking" in a single instrument of investigation the different tools developed for investigation in each of the different areas.

2.1. Complexity Of The Approach

The scope is the safety of nuclear power plants, is demonstration of the capability to keep the radiation exposure of personal and population within specified limits. It is ensured by maintaining the integrity of safety barriers, which are part of the plant defence in depth concept.

A series of barriers prevents the release of radioactive fission products from their source beyond the reactor containment and into the environment. In analyzing the NPP safety, it is essential to assess the integrity of these barriers and to decide to what degree the response of the whole NPP and its systems to a certain initiating event is acceptable from the viewpoint of the plant safety. The integrity of the safety barriers is related to certain threshold values, which are referred to as acceptance criteria. Design limits are adopted with a conservative margin so that the safety barrier integrity is guaranteed as long as the parameters do not exceed the relevant criteria. In the case of not efficacy of the barriers a radioactive release occurs and an evaluation of the dose to the workers and population is done (IAEA, 1996and IAEA, 2000).

The complexity of the analysis is due to the involvement of a number of different technological areas requests a detailed identification of topics and targets together with a suitable connection with adopted codes and activities.

The nuclear technology sectors or computational areas relevant for NPP safety and design include the following areas: the system thermal-hydraulics, the computational fluid-dynamics, the structural mechanics, the neutron kinetics with the cross section generation and the fission product release and transport.

The interconnections among individual Technological Areas identify a chains of codes. Figure 1 gives an idea of the complexity of the activities and related technological areas necessary for a such analysis.

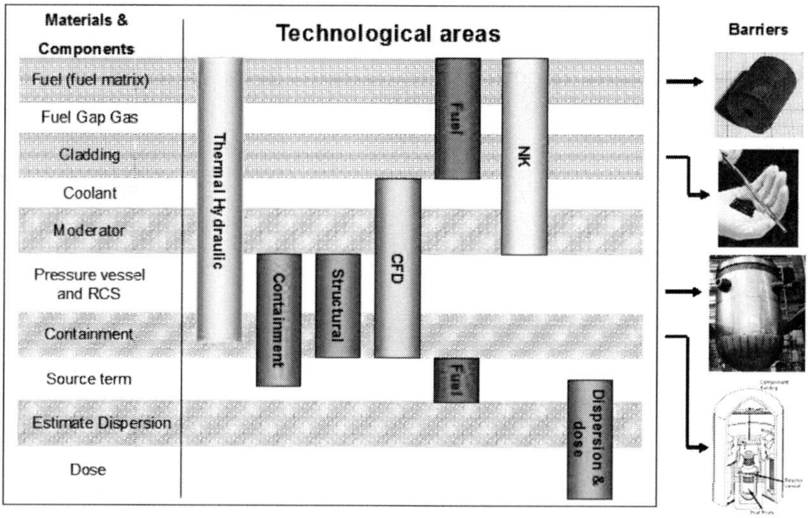

Figure 1. Technological areas for the integrated an analysis

The effort to perform a such analysis is aimed to establish a connection with the regulatory or licensing environment. This connection must take into account the evolution of safety concepts following improvements of the technical knowledge, including the availability of powerful computational tools and of experimental evidences.

The framework constitutes by the development & qualification of computational tools is also related to relevant points like "physical phenomena understanding", and "analysis of complex scenarios expected during accident conditions" considering the current licensing practices.

The strategic objective is the set-up of a suitable chain of codes to deal with accident scenarios. The motivation for the selection of individual accidents is given by expecting challenging phenomena for the concerned safety barrier. The concerned phenomena shall also be connected with the existing code typologies and capabilities. These codes are supposed to be qualified for the prediction of individual accidents whose relevant and detailed boundary and initial conditions have been defined.

The list of phenomena, which are taking place during progression of an accident shall be analyzed, discussed and selected. Relevant information can be taken from international literature (e.g. IAEA, 2002) or from experimental tests. The operative objective is to demonstrate the capability of computational tools to reproduce relevant transient phenomena and to show that the same tools can be linked together.

Generally speaking, best estimate is associated to the TH SYS codes. About this kind of codes is clear the meaning of best estimate approach. Descriptions of this concept are largely diffused in international literature. The concept of best estimate is less clear about the codes related to the other technological areas. The general concept of best estimate approach is in avoiding any intentional conservatism. This concept is applied in all the aspects of the calculation: input data, conditions of the calculation, model of the systems and of course the code. From this point of view the aspects to be considered for each individual codes are:

- The physical modelling.
- The approximations that are made and their limitations.
- The used correlations.
- An assessment of uncertainties due to the physical models.
- The practice of application associated to these codes
- and their level of validation and/or certification.
- the associated impact on the drawing of safety analyses.

In such a complex analysis, requiring different codes, the data used as input for a code are derived from the result of another previous code calculation. So a relevant role is also played by the evaluation and selection (as input data in next calculations) of the results obtained by code application. The figure. 2gives an idea of the links between the different technical areas.

Referring to the figure 2, some links are hereafter exemplified.

- Path a) the TH codes results are used to supply boundary data to the code for fuel evaluation.
- Path b) the TH codes supply the thermal hydraulic boundary conditions to the NK codes. The results of the NK codes are supplied to the TH code core component.

- Path c) the TH codes supply the thermal hydraulic boundary conditions to the CFD codes. The results of the CFD codes are supplied to the TH for evaluation of specific areas of the systems.

- Path d) the CFD codes supply the boundary conditions to the Structural code for evaluation of mechanical resistance of systems components.

- Path e) the results from the Containment code are supplied to the TH codes for calculation of the evolution of the accident in the reactor coolant system and containment.

- Path f) the results of the Structural code about the integrity of the systems (e.g. containment systems) are supplied to the containment codes.

- Path g) the results of the Containment code about possible failure and source terms are supplied to the codes for dispersion and dove evaluation.

- Path h) the results of the Fuel code about source terms are supplied to the codes for dispersion and dove evaluation.

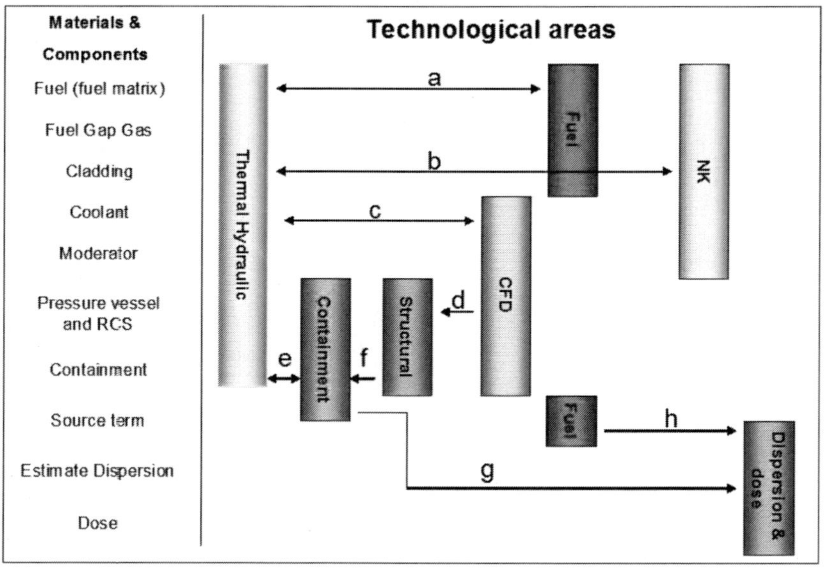

Figure 2. Links between the different technical areas

2.2. Qualification and Uncertainty

A relevant aspect in best estimate application is the qualification of the process of code application:

The following specific topics must be covered:

- Development process of generic codes and their capabilities;
- Developmental Assessment;
- Structure of specific codes
- Numerical methods;
- Description of input decks;
- Description of fundamental analytical problems;
- Analysis of fundamental problems;
- International Standard Problem Activity and benchmarks;
- Example of code results from applications to ITF;
- Plant accident and transient analyses application;
- Modalities for developing the nodalization;
- Description and use of nodalization qualification criteria;
- Qualitative and quantitative accuracy evaluation;
- Use of thresholds for the acceptability of results for the reference case;
- Description of the available uncertainty methodologies;
- Coupling methodologies.

A specific aspect of best estimate application is constitute by uncertainty evaluation (Wickett et al., 1998). For the TH codes specific methodologies were developed and applied. International literature offers a spread documentation about this uncertainty methodologies for TH codes.

Concerning the codes not connected with the TH area the following items must be evalauted to derive the evaluation of the uncertainty.

- Description of the numerical methods. Generally the codes are validated versus some reference calculations and the related uncertainty is also given.
- International Standard Problem Activity and benchmarks. From the comparison with the result of other qualified codes can be estimated the uncertainty of the code.
- Code application to experimental tests.
- Code application to experimental tests in Plant accident and transient analyses.

Additional and relevant aspects to be also considered are:

- Procedure for developing the nodalization developed by the user or in the code manual.
- Description and use of nodalization qualification criteria.
- User experience

2.3. Computational Tools Needed In The Analysis

The computational tools include:

- the best estimate computer codes;
- the nodalization including the procedures for the development and the qualification;
- the uncertainty methodology including the procedure for the qualification;
- the computational platforms for coupling and interfacing inputs and outputs from the concerned codes and nodalization.

An outline of the codes listed in the table below is provided in the table 1.

All considered codes should be well established within the international community and some referenced document per each code should be provided that gives access to the peculiarities of the code.

Table 1. Outline of the codes needed in the analysis

No	Field of application	Example of applications
1.	System Thermal-Hydraulics	All transients
2.	I&C Modelling	All transients (where I & C, i.e. control, limitation and protection systems, play a role).
3.	Computation Fluid Dynamics	Special detailed analyses of specific components and/or systems
4.	Structural Mechanics	PTS and structural mechanics integrity of the vessel wall.
5.	Fuel (mechanics)	All transients in relation to which the number of failed rods is calculated
6.	Neutron Physics (and supporting)	Transients analyzed by 3D coupled neutron kinetics - thermal-hydraulics: spatial or local neutron flux effects are relevant – transient conditions.
	Confinement	Severe accident
7.	Radiological Consequences (and supporting)	Environment diffusion and dose tot the population

Key issues for the application of the codes are represented by:

a) the demonstration of the code qualification level;

b) the demonstration of the current user capabilities in the use of the codes.

The quality demonstration of individual codes, item a), can be derived by several hundred worldwide available documents. In addition to such documents, per each code there are specific-additional qualification documents issued. The reference document provided per each code, gives one access to international qualification documents.

Connected with the above item a), the quality of the code application results is increased by a systematic and comprehensive application of independent codes for deriving the same result. All the codes should be applied by the users, item b), having experience (years) in the code application and results analysis. Code qualification cases shall be considered in order to prove the user capabilities in the application of the codes.

2.3.1. System Thermal-Hydraulics

The quantitative characterization of a system transient scenario constitutes the main role for the System Thermal-hydraulic (SYS-TH) code, consistently with the main objective for its development. The SYS-TH code gives the results connected with the thermal hydraulic parameters evolution of the NPP during a transient. The application of the SYS-TH code, because of the capability to represent all the systems in a quit compact and fast calculations, is typically also used to derive the initial conditions for the application of other more specific codes/tools.

These kinds of codes generally have embedded some additional capabilities:

- The multi-dimensional component in SYS-TH code developed to allow the user to more accurately model the multi-dimensional flow behaviour that can be exhibited in any component or region of a system.
- Neutron kinetic modules: the NK module can have from zero to three dimensions representation capabilities.
- Severe accident module: a limited capability can be included in simulating core damage occurrence and fission fragment distribution in the systems.

2.3.2. I&C Modeling

The aim is to simulate the performance of the control, the limitation and the protection systems of the NPP. The simplified representation of the protection system only could be not sufficient for a detailed analysis. The Instrumentation and Control (I&C) can be modelled in the SYS-TH code. But the complexity of the control (also including limitation) systems request a more capable end flexible tool. Some applications have been done just realizing software (e.g. Fortran based software) coupled with the SYS-TH code.

In the I&C software the equations are solved to simulate the transient behaviour of the various transducers, actuators and logic of operation of each individual component that constitutes the control, the limitation and the protection systems of the NPP. The code receives the system information at each time step from the SYS-TH code related to any requested thermal-hydraulic variable (e.g. pressure, level, pressure drop, fluid temperature). The related information is processed, e.g. considering the inertia of the transducer or the delay of the signal transmission, and commands for components (typically pumps, valves, control rods, heaters, etc.) modelled in SYS-TH are generated. With the new system configuration a new time step is calculated and the above process starts again.

2.3.3. Computational Fluid Dynamics

The main role of CFD is to support and validate the application of the SYS-TH in relation to the mixing phenomena and in calculating pressure drop coefficients at geometric discontinuities where information from experimental data is not adequate. The latter role is also relevant to the PTS study.

CFD features the following modelling capabilities:

- Steady-state and transient flows.
- Laminar and turbulent flows.
- Subsonic, transonic and supersonic flows.
- Heat transfer and thermal radiation.
- Buoyancy.
- Non-Newtonian flows.
- Transport of non-reacting scalar components.
- Multiphase flows.
- Combustion.
- Flows in multiple frames of reference.
- Particle tracking.

2.3.4. Structural Mechanics

The structural mechanics code is used to calculate stress and strains in components other than the fuel rods. Two main uses are exemplified in order to summarize the role of the code:

a) demonstration that dynamic loads, following transient scenarios, do not cause rupture/collapse of or the substantial deformation of the relevant component potentially affecting the coolability of the core;

b) calculation of stresses in the components relevant to prevent radioactive releases. Typical application is constituted by PTS analysis.

These tools are adopted to perform static and dynamic analyses of linear and non-linear problems (due to materials properties, geometry, contact between surface, etc.) in many fields of application (structural, thermal, electromagnetic, fluid-dynamic, etc.). It is possible to solve coupled problems as well as fluid–structure interaction, thermal–mechanical calculation. In addition several special purpose features are available, namely: fracture mechanics, composites, fatigue, beam analyses.

2.3.5. Fuel Mechanics

The key goal for the use of the code is the evaluation of the integrity of the fuel claddings. The number of nuclear fuel rod claddings that are damaged following each transient constitutes the typical output from the code. The code is a computer program for the thermal and mechanical analysis of fuel rods in nuclear reactors. The code was specifically designed for the analysis of a whole rod. Code incorporates physical models of thermal and radiation densification of the fuel, models of fuel swelling, fuel cracking and relocation, a model of generation of fission gases, a model of redistribution of oxygen and plutonium, and some other physical models. The code has the capabilities of analysis of all fuel rod types under normal, off-normal and accident conditions (deterministic and probabilistic).

2.3.6. Neutron Physics

The transient (time dependent) three-dimensional calculation of the neutron flux following global or local perturbations constitutes the main goal fro the use of the code. The neutron kinetics subroutines require as input the neutron cross-sections in the computational nodes of the kinetics mesh. A neutron cross-section model has been implemented that allows the neutron cross-sections to be parameterized as functions of SYS-TH code heat structure temperatures, fluid void fraction or fluid density, poison concentration, and fluid temperatures. Additional codes are necessary to (not exhaustive list):

- to derive macroscopic cross sections thus supporting the application of the Nestle code;
- to support and to validate calculation results (fluxes and several reaction rates in each point of the calculation domain and to perform criticality analyses);
- to calculate fuel cell calculation versus burn-up;

- to calculate the build up, decay, and processing of radioactive materials;
- to convert evaluated nuclear data file in continuous-energy or multi-group microscopic cross sections libraries.

2.3.7. Radiological Consequences

The purpose is to simulate the impact of severe accidents at nuclear power plants on the surrounding environment. The principal phenomena considered are atmospheric transport, mitigation actions based on dose projections, dose accumulation by a number of pathways including food and water ingestion, early and latent health effects, and economic costs.

Several aspects must considered:

- Calculation of the radioactivity inventory in the fuel elements.
- Tracking the transport of radioactivity products inside the primary system and the containment.
- Calculating the offsite radioactivity dispersion and the dose to the population.
- Calculating the onsite dispersion and the dose to the control room personnel

2.3.8. Nodalizations

The nodalizations are the result of a brainstorming process by the code-users, which connect each code with the physical system to be simulated. The process for developing a nodalization especially for a best estimate code does not necessarily require less effort than the process of development of the code itself. The same is true in relation to the qualification. Expert users develop the nodalization for an assigned purpose, provided that Best Practice Guidelines are followed whenever available. Sensitivity tests can be performed to demonstrate the nodalization quality and the achievement of mesh-independence of the results, which means that varying the node density (or the number of nodes) does not make the results change to a large extent. All nodalizations shall be developed according to suitable quality assurance procedures and criteria. The procedures are linked with the code characteristics and with the expertise of the users.

All nodalizations developed to apply the BE codes must be qualified according to current standards that are specific for each code. Plant nodalization should be developed according to predefined qualitative and quantitative acceptance criteria.

Three major steps in the process must be distinguished each one characterized by a number of sub-steps, by procedures and by acceptability thresholds:

1 Nodalization development: the nodalization must be characterized by 'geometric fidelity' with the modelled physical systems that are part of the NPP.

2 Acceptance of steady state.

3 The transient capability: the capability of the code-nodalization in simulating the phenomena of interest must be demonstrated

Qualitative and quantitative acceptability thresholds and criteria are adopted at step 1). Quantitative acceptability thresholds are adopted at step 2). Qualitative and quantitative accuracy evaluation is performed for step 3) with quantitative thresholds.

A simplified scheme of a procedure for the qualification of the nodalization is depicted in the figure 3. It is assumed that the code has fulfilled the validation and qualification process and a "frozen" version of the code has been made available to the final user. The steps of the diagram are described below.

3. EXAMPLE OF APPLICATION: INTRODUCTION TO THE ANALYSIS OF THE MPTR

The RBMK core is constituted by more than one-thousand pressurized channels housed into stacked graphite blocks and connected at the bottom and at the top by small diameter (D) and long length (L) pipes (less than 0.01 and more than 10 m, respectively) that end up into headers and drum separators. Control valves are installed in the bottom lines. Due to the large L/D value and to the presence of valves and other geometric discontinuities along the lines connecting with the pressure channels, the Fuel Channel Blockage (FCB) event is possible and already occurred in two documented NPP events. Previous investigations, have shown the relevance of these events for the safety technology, and the availability of proper computational technique for the analysis (NIKIET, 1983and 1992).

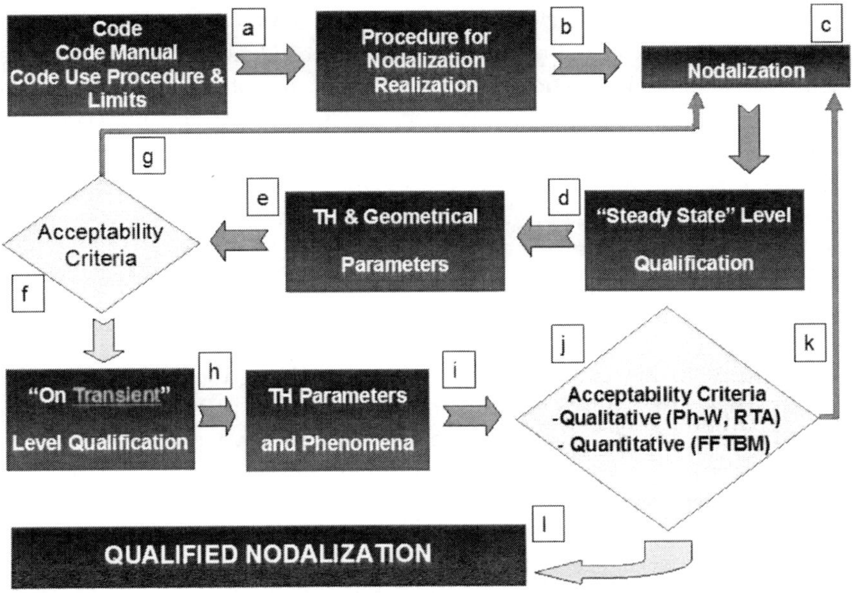

Figure 3. Simplified scheme for nodalization qualificationStep "a":this step is related to the information available by the user manual and by the guidelines for the use of the code.Step "b": user experience and developers recommendations are listed and considered.Step "c": the nodalization must reproduce all the relevant parts of the reference plant; this includes geometrical and materials fidelity and consideration of components and logics.Step "d": different checks are performed under this step mostly geometry related (does not require running the code-nodalization).Step "e": different checks are performed under this step.Step "f": this is the step where the adopted acceptability criteria are applied to evaluate the comparison between hardware and implemented geometrical values in the nodalization and between the experimental and calculated steady-state parameters.Step "g": if one of the criteria in the step "f" are not fulfilled, a review of the nodalization (step "c") must be performed. The path "g" must be repeated till all acceptability criteria are satisfied. Step "h": this step constitutes the "On Transient" level qualification and allows the verification of selected data that are relevant only during transient.Step "i": in this step the thermal-hydraulic parameters that are at the basis of the qualitative or quantitative accuracy evaluations are characterized.Step "j": checks are performed to evaluate the acceptability of the calculation, e.g. of the 'Kv-scaled' calculation both from qualitative and from quantitative points of view. Step "k": this path is actuated if any of the checks (qualitative and quantitative) is not fulfilled. Step "l": the obtained nodalization is used for the selected transient and the selected facility or plant. Any subsequent modification of the nodalization requires a new qualification process both at "steady state" and at "on transient" level.

The occurrence of the FCB event remains undetected for a few tens of seconds because of the lack of full monitoring for the individual channels. Therefore, fission power continues to be produced in the absence of cooling. This brings in subsequent times to fuel rod overheating, pressure tube failure, damage of the neighbouring graphite brick and ejection of damaged fuel. Following the pressure tube rupture, reactor cavity pressurization, radioactivity release into the same area and change of fluid properties occur that allow the detection of the event and cause the reactor scram at a time of a few tens of seconds depending upon the channel working conditions and the severity of the blockage.

Notwithstanding the scram and the full capability of the reactor designed safety features to keep cooled the core, the multiple pressure tube rupture (MPTR) issue is raised. The question to be answered is whether the 'explosion' of the blocked pressure tube damages not only the neighbour graphite bricks but propagates to other channels causing the potential for several channel failure.

In order to address the MPTR issue fuel channel thermal-hydraulics and three-dimensional (3D) neutron kinetics analyses have been performed, as well structural mechanics calculations for the graphite bricks and rings (graphite rings surround the pressure tube to accommodate for thermal and radiation induced expansions).

The bases for the analysis and the results of the study are presented. The conclusion, not reported within a licensing based format, is that the MPTR consequences are not expected to be relevant for the safety of the RBMK installations.

3.1. Execution Of The Analysis

The detailed knowledge of the RBMK system configuration was not spread in the Western world till the 1986 event. Afterwards, "information batches" of RBMK technology became available and were unavoidably evaluated in the light of the Chernobyl event. The results of recently completed project sponsored by European Commission (EC), with the participation of RBMK designers in Russia and the supervision of the national utility and the regulatory authority, allow to give an idea of RBMK current safety characteristics. The project has been made possible owing to the availability of sophisticate computational tools developed and qualified in the last decade. These include powerful computers, advanced numerical solution methods, techniques for developing input decks and for proving the qualification level. Following the identification and the characterization of bounding scenarios assuming to envelope all accident conditions relevant to RBMK safety technology, two main chains of codes have been set-up and utilized to perform safety analyses.

3.2. The Computational Tools

The computational tools include the numerical codes, the nodalizations and the relevant boundary and initial conditions related to the Smolensk 3 NPP in the present case. The application of computational tools requires systematic demonstration of quality and suitable documentation detail. However, within the scope of the performed activity, there is the 'as-far-as-possible' demonstration of quality for codes, the development of nodalizations, the implementation of boundary and initial conditions as available and the achievement of results from computer calculations. Furthermore, terms like 'capable code' and 'suitable code' have been introduced. A code is 'capable' when it is able to simulate the phenomena and the physical scenarios expected during the assigned NPP accident. A code is 'suitable' when a user can run the code addressing (or calculating) the expected phenomena within a reasonable time with reasonable resources. It should be noted that the term 'capable' is less binding for a code than the term 'qualified' and a quantification is provided for the items 'reasonable resources' and 'reasonable time'.

3.3. The Numerical Codes

The numerical codes adopted are those listed in the third column of Table 2.

The area for the application of the codes can be deduced from the second column in the same table and from the diagrams in figure. 4 and figure 5 that are applicable for the Russian and the Western codes, respectively. Topological subjects relevant to the deterministic safety analysis of RBMK are identified in Figs. 4 and 5 and the correspondence with the range of application of numerical codes is established.

Table 2. Adopted numerical codes

Identification		Codes adopted	Reasons for the selection
No.	**ACRONYM explanation**		
A1	LOCA-PH-FIGDH: LOCA in Pressure Header with failure to isolate GDH	Relap5	Largest primary system break with single failure. Challenging core cooling and the ECCS design
A2	LOCA-SL: LOCA originated by a break in Steam Line		Highest depressurization rate. Challenging core cooling and the ECCS design
A3	LOOP-ATWS: Loss of on Site Power with the ATWS condition		Challenging core cooling and the neutron kinetics model of the thermal-hydraulic system codes
A4	GDH-BLOCKAGE: Full blockage of the GDH		Check of the capability of the 'ECCS bypass' to cool the core
B1	GDH-BLOCKAGE-SA: Full blockage of the GDH with the 'Severe Accident' assumption of no bypass line available	Cocosys and Relap5	Challenging the venting capability of the reactor cavity (part of the confinement)
B2	LOCA-PH-FIGDH: See A1	Contain and Relap5	Challenging the ALS (part of the confinement) structural resistance (same as A1)
B3	LOCA-SL: See A2	Contain	Challenging the reactor building (part of the confinement) venting capability (same as A2)
C1	FC-BLOCKAGE: Full blockage of one fuel channel	Relap5-3D©/Nestle	Challenging the calculation of the local fission power generation (same as D1)
C2	GDH-BLOCKAGE: See A4		To assess and to understand the local core response (same as A4)
C3	CR-G-WITHDRAWAL: Continued withdrawal of a CR bank (or group)	Korsar-Bars	Challenging RIA (Reactivity Initiating Event)
C4	CPS-LOCA: Voiding (or LOCA) of the CPS	Relap5-3D©/Nestle	
D1	FC-BLOCKAGE: See C1	Relap5-Ansys Katran-U-Stack	Driving accident for the study. Challenging various areas and codes
D2	FC-LOCA: Rupture of one FC	Contain & Relap5 Fluent-Ansys Korsar-Rapta	To assess the ballooning model in the fuel pin mechanics area
E1	FC-BLOCKAGE: See C1	Cocosys Melcor	To assess the hydrogen and the fission products source term and transport (same as B1)
E2	GDH-BLOCKAGE-SA: See B1		To assess the hydrogen and the fission products source term and transport in one extreme conditions (same as B1)
F1	FC-BLOCKAGE: See C1	Relap5	To formulate the ICM proposal (same as D1)

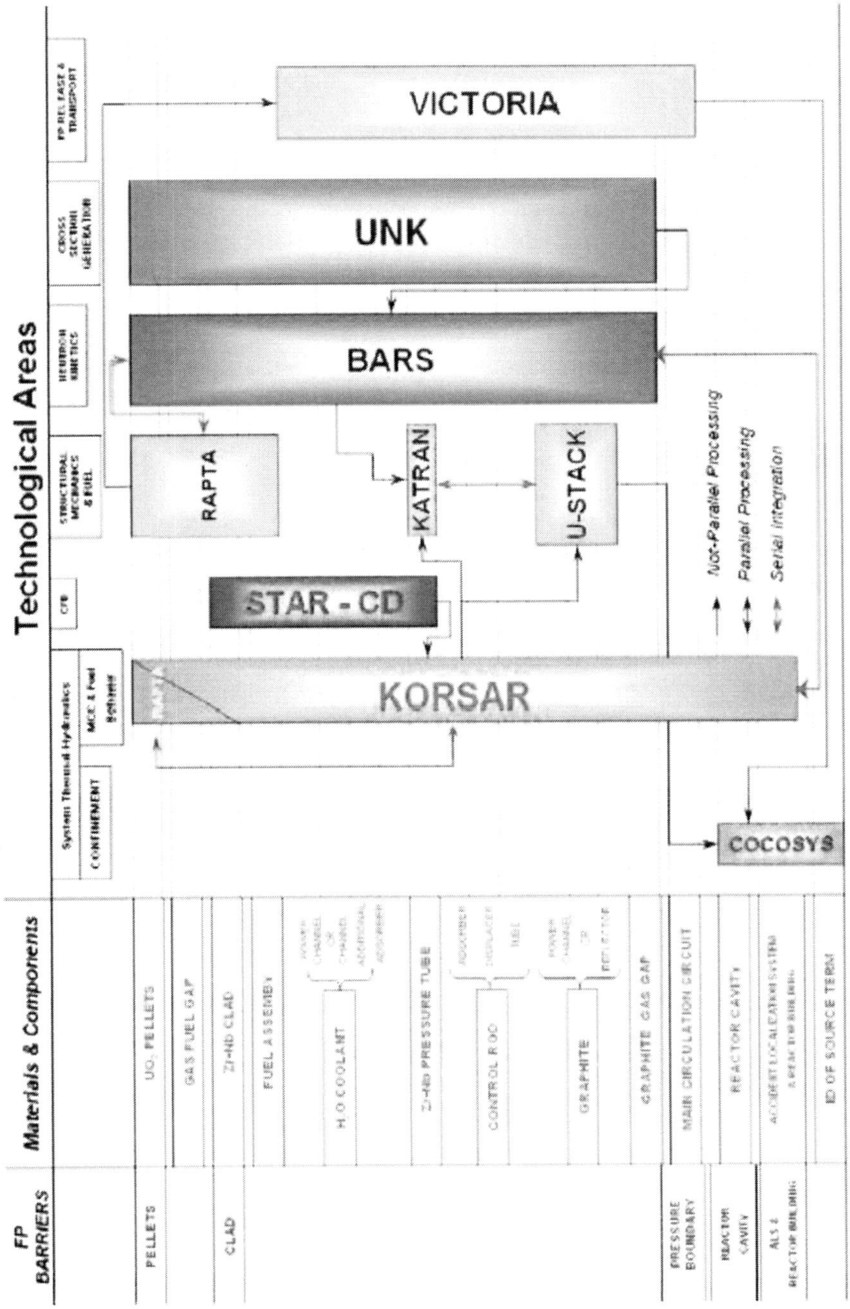

Figure 4. Codes adopted by Russian group

The topological subjects include:

- Five fission product barriers: the fuel pellet, the clad, the pressure boundary of the primary cooling system and the confinement regions corresponding to the reactor cavity, the (ALS) and the reactor building.
- The materials and components constituting the NPP hardware: the coolant, the fuel and the moderator are examples of 'materials'; the control rods, the pressure tube and the zones of the confinement are examples of 'components'.

The technological areas (for deterministic safety analysis) include the system thermal-hydraulics, the computational fluid-dynamics, the structural mechanics, the neutron kinetics with the cross section generation and the fission product release and transport.

3.4. The Nodalizations

Nodalizations were developed for both Western and Russian codes by modelling the materials and components, by making reference to the technological areas and by considering the features of codes with the target of demonstrating codes capability and suitability, but also to assess the integrity of the fission product barriers. Nodalizations are typically the result of wide range brainstorming processes whose outcome depends upon the code features, the available computer power, the expertise of the user and the target for the analyses. An example of the realized nodalizations is reported in the table 3.

3.5. The Boundary and The Initial Conditions

Boundary conditions for NPP accident analyses are constituted by huge amount of data ranging from in the present case the mass of water in the steam drum, to the individual fuel bundle burn-up, to the material properties of irradiated graphite, to the thickness and the Young module for the tank that encompasses the graphite stacks, to the free volume of the reactor cavity, to the net flow areas of the valves/openings connecting various zones of the confinement with the environment.

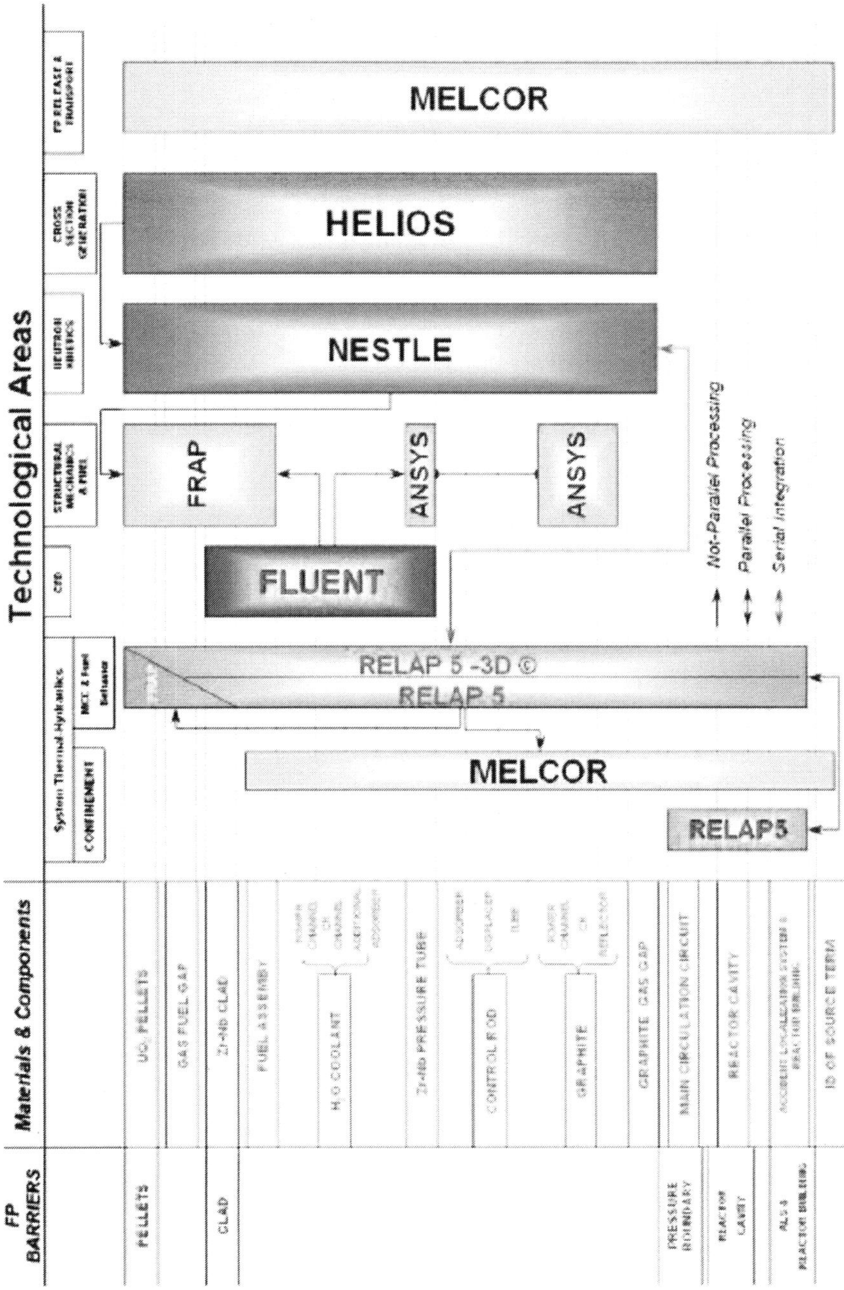

Figure 5. Codes adopted by western group

The boundary conditions for the MPTR issue is the accident scenario originated by the fuel channel blockage (FC-BLOCKAGE making reference to boundary conditions in the Smolensk-3 NPP unit.

3.6. THE MULTIDISCIPLINARY PROBLEM ASSOCIATED WITH THE FC-BLOCKAGE SCENARIO

The background for addressing the multidisciplinary problem arising from the FC-BLOCKAGE and the MPTR include the presentation of following aspects:

- study of the components and zones of the RBMK core region to make clear the concerned accident scenario,
- the characterization of the steady state operation of the reference RBMK boiling channel,
- the experience from the pressure tube (PT) rupture events in RBMK NPP,
- the phenomenological evolution of the transient.

An overview of the multidisciplinary problem associated with the blockage of one fuel channel scenario is given by the phenomenological aspects associated with the scenario originated by the blockage of one fuel channel in the RBMK NPP (i.e. FC-BLOCKAGE event). To this aim, phenomena are identified that characterize the progression of the event, the failure map for RBMK pressure tubes and the probable position for break elevation following FCBLOCKAGE.

The multidisciplinary nature and the demonstration of complexity for the concerned scenario is shortly highlighted in the figure 6 e table 4 and can be summarized in the following list:

- System Thermal-Hydraulics related to reactor coolant system
- Fuel pin mechanics to evaluate the fuel performance parameters including rod deformation following the FC-BLOCKAGE event in the RBMK fuel bundle.
- System Thermal-Hydraulics related to confinement
- Computational fluid dynamics to calculate the hydraulic loads acting upon the fuel rods following the rupture of the pressure tube occurring during the FC-BLOCKAGE event.
- Neutron kinetics for generation of average parameters for microscopic cross-sections as a functions of energy

Table 3. Realized nodalizations

No.	System or component	Code	Total number of nodes	Target for the nodalization
1	Main coolant circuit	Relap5	6491/52735*	Thermal-hydraulic performance of the RBMK primary loop in case of DBA and BDBA
2		Korsar	829/5256*	
3	Confinement system	Relap5	621/2625*	Thermal-hydraulic performance of the RBMK confinement in case of DBA and BDBA
4		Cocosys	37	
5	Reactor core	Nestle	29856	Three-dimensional calculation of fission power in the core
6		Bars	17416	
7	Part of a FC	Fluent	$\sim 10^6$	Calculation of fluid-dynamic forces acting upon fuel rods following FC break
8	FC Inlet Zone	Star-cd	$3.7\ 10^5$	Calculation of pressure and velocity fields to assess the appearance of steam cavitation in control valve
9	Broken Single Channel and neighboring FC	Relap5	167/2810*	Calculation of break area and position and 'local' hydraulic parameters for primary system and confinement
10	Individual Pressure Tube and connected Graphite Stack	Ansys	4700	Stresses in pressure tube and in graphite stack consequent to the creep deformation of pressure tube
11	Fuel moderator and coolant	Helios	$14 \times 4 + 30 \times 4$ (zones)	To derive macroscopic cross section sets suitable for the use of three-dimensional neutron kinetics codes
12		MCNP	Not applicable	
13	Graphite stack as a whole	U_Stack	–	To address the MPTR issue by the deformation method

* Number of hydraulic nodes and number of mesh points for conduction heat transfer are distinguished.

- Neutron kinetics: 3D transient neutron flux to calculate the neutron kinetics parameters in the individual fuel channel and associated graphite stack following the FC-BLOCKAGE event.

- Structural mechanics to calculate stresses and strains in the pressure tube and in the graphite blocks following the rupture of the pressure tube occurring during the FC-BLOCKAGE event.

- Fission products generation to calculate the source term associated with the operation of a fuel channel of the RBMK, i.e. the amount of radioactivity that is released during the progression of the FC-BLOCKAGE event.

- Fission products transport to calculate the transport of the fission products generated as a consequence of the melting and the damage of a RBMK fuel bundle during the progression of the FC-BLOCKAGE event.

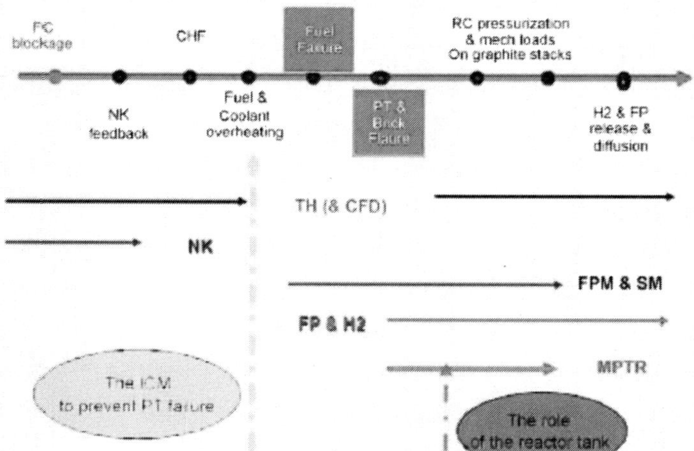

Nomenclature: *CFD = Computational Fluid Dynamics, CHF = Critical Heat Flux, FC = Fuel Channel, FP = Fission Products, FPM = Fuel Pin Mechanics, ICM = Individual Channel Monitoring (see D'Auria et al. 2007e), MPTR = Multiple Pressure Tube Rupture, NK = Neutron Kinetics, PT = Pressure Tube, RC = Reactor Cavity, SM = Structural Mechanics, TH = Thermal-hydraulics.*

Figure 6. Scheme of the multidisciplinary approach

Table 4. Technical areas involved in the analysis

No.	Technological safety area		Codes	Notes
1	System Thermal-Hydraulics	Primary System	Relap5, Korsar	Also includes H2 generation and transport
2		Fuel	Frap, Rapta	
3		Confinement	Cocosys, Relap5	
4	Computational Fluid Dynamics		Fluent	
5	Structural Mechanics		Katran, Ansys (U_Stack)[a]	The MPTR issue is addressed by the U_Stack code and by a procedure (given in chapter 6)
6	Neutron Kinetics	Generation of average parameters (e.g., macroscopic cross-sections or 'λ-functions')	Njoy, Unk, Helios	
7		3D transient neutron flux	Bars, Nestle	
8	Fission Products	Generation	Refp, Melcor	
9		Transport	Cocosys, Melcor	

[a] U_Stack code has capabilities to handle technological safety areas 1, 2, 3 and 5.

3.7. Results Of The Analysis

The scenario puts an enormous challenge to the codes: all key technological areas relevant to the deterministic reactor safety are involved. About 40 phenomena have been identified as characterizing the scenario and related computational tools have been evaluated.

However the possibility for the occurrence of the multiple pressure tube rupture (MPTR) was excluded.

4. CONCLUSIONS

The best estimate concept is defined as the efforts in avoiding conservative assumptions in performing analysis. It implies to adopt the best suitable tool available for each specific topic relevant for an analysis. In the case of an analysis related to a like NPP complex system it is necessary to "enlarge" the investigation in many technological areas. A direct consequence is constituted by the adoption of an integrated approach in performing safety analysis. A further relevant consequence is that the best estimate concept must be applied to a broad spectrum of disciplines.

This integrated best estimate approach for safety analysis means the availability of qualified tools an qualified users in many technical areas. The qualification has to be also applied to the coupling of the codes typically organized in a sort of "chain" including not only the code itself but also the input of the codes and the input and output data.

It is also relevant the availability of a suitable computational power necessary to perform the calculation with the different codes. The importance of this aspect is connected with the capability to include in the calculations all the details necessary to obtain a results to be considered as best estimate input for other linked codes.

The uncertainty is another relevant aspects. Best estimate application always requests uncertainty evaluation. The uncertainty evaluation is rather well developed for TH SYS code, but requires a focused and special effort in the case of all the other technical areas.

Summarizing, the best estimate concept applied in analysis for complex systems should be applied as an integrated approach in the meaning of application covering many technological areas and it requests a large effort in terms of technical competences, capability in qualified use of tools and user, and computational power.

REFERENCES

1. International Atomic Energy Agency 2008 Best Estimate Safety Analysis for Nuclear Power Plants: Uncertainty Evaluation, Safety Reports Series 52 IAEA 978-9-20108-907-6 Vienna

2. International Atomic Energy Agency 1996 Defense in Depth in Nuclear Safety, INSAG-10, IAEA 9-20102-596-3

3. International Atomic Energy Agency 2000 Safety of Nuclear Power Plants: Design, Safety Standard Series No. NS-R-1, IAEA 9-20101-900-9

4. International Atomic Energy Agency 2002 Accident Analysis for Nuclear Power Plants, Safety Reports Series 23 IAEA 9-20115-602-2

5. NIKIET 1992 Accident Analysis of rupture of fuel channel 5216 at Leningrad NPP Unit 3 (in Russian). Internal Report 040-116-4101, Moscow (Ru)

6. NIKIET, 1983 Accident Analysis of 6244 Fuel Channel Rupture at Chernobyl-1: Reasons and Consequences (in Russian). Internal Report 040-103-1571d, Moscow (Ru).

7. T. . Wickett, Editor, F. D'Auria, H. Glaeser, E. Chojnacki, C. . Lage, Authors. Lead, D. Sweet, A. Neil, G. M. Galassi, S. Belsito, M. Ingegneri, P. Gatta, T. Skorek, E. Hofer, M. Kloos, M. Ounsy, J. I. Sanchez, 1998 Report of the Uncertainty Method Study for Advanced Best Estimate Thermal-hydraulic Code Applications- Vols. I and II OECD/CSNI Report NEA/CSNI R (97) 35, Paris (F)

Chapter 2

Design process of the nanofluid injection mechanism in nuclear power plants

Myoung-suk Kang[1], Changhyun Jee[1], Sangjun Park[1], In Choel Bang[2] and Gyunyoung Heo[1*]

1Kyung Hee University, 1 Seocheon-dong, Giheung-gu, Yongin-si, Gyeonggido 446-701, Republic of Korea

2Ulsan National Institute of Science and Technology, 100 Banyeon-ri, Eonyang-eup, Ulju-gun, Ulsan-si 689-798, Republic of Korea

ABSTRACT

Nanofluids, which are engineered suspensions of nanoparticles in a solvent such as water, have been found to show enhanced coolant properties such as higher critical heat flux and surface wettability at modest concentrations, which is a useful characteristic in nuclear power plants (NPPs). This study attempted to provide an example of engineering applications in NPPs using nanofluid technology. From these motivations, the conceptual designs of the emergency core cooling systems (ECCSs) assisted by nanofluid injection mechanism were proposed after following a design framework to develop complex engineering systems. We focused on the analysis of functional requirements for integrating the conventional ECCSs and nanofluid injection mechanism without loss of performance and reliability. Three candidates of nanofluid-engineered ECCS proposed in previous researches were investigated by applying axiomatic design (AD) in the manner of reverse engineering and it enabled to identify the compatibility of functional requirements and potential design vulnerabilities. The methods to enhance such vulnerabilities were referred from TRIZ and concretized for the ECCS of the Korean nuclear power plant. The results show a method to decouple the ECCS designs with the installation of a separate nanofluids injection tank adjacent to the safety injection tanks such that a low pH environment for nanofluids can be maintained at atmospheric pressure which is favorable for their injection in passive manner.

1. INTRODUCTION

One of the methods for enhancing the safety of nuclear power plants (NPPs) is related to improve their heat transfer capability. Nanofluids are known to exhibit superior heat transferability and are therefore being actively investigated for engineering applications [1–4]. Recently, the studies on the introduction of nanofluids for emergency core cooling systems (ECCSs) which is one of engineered safety features of NPPs were conducted [5–7]. Such works were characterized by the enhancement of critical heat flux (CHF) via nanofluid injection in cases of loss of coolant accidents (LOCAs). While taking the benefits of nanofluid injection during accident conditions, it is apparent that the nanofluid-engineered ECCSs should be compatible with conventional systems during normal operations to make nanofluid technologies practical in NPPs. With this motivation, it is, therefore, important to analyze the functional requirements (FRs) for integrating the ECCS and nanofluid injection mechanism without loss of performance and reliability of the conventional systems.

This study employs axiomatic design (AD) and TRIZ for analysis of FRs and creation of a relevant nanofluid-engineered ECCS. The theory underlying AD is based on the hypothesis that superior design begins with certain axioms that facilitate the creation of systems through interactive mapping of FRs and design parameters (DPs). In addition, AD facilitates reasonable and logical steps that take conceptual design processes into account [8, 9]. The second tool employed in this study TRIZ, which is a Romanized acronym of a Russian phrase meaning, 'theory of solving inventor's problems' or 'theory of inventive problem solving' [10]. The TRIZ theory applies abstraction and concretization processes to facilitate the creation of solutions for problems recognized from an existing design by the AD process [11].

In this study, the analysis of a conventional design of an ECCS and several design alternatives to adopt nanofluid injection mechanism using the principles of AD has been presented, and also the compatibility of FRs and potential design vulnerabilities are discussed. The methods to enhance such vulnerabilities are drawn from TRIZ and finally concretized a specific design of a nanofluid-engineered ECCS. In this article, a conceptual design incorporating nanofluids for the ECCSs of the Korean Advanced Power Reactor 1400MWe (APR1400) has been elicited.

2. BACKGROUND

2.1 LOCA and ECCS

A LOCA is an accident which occurs due to a break in a reactor coolant system (RCS) pipeline in a NPP. LOCAs are considered to be serious accidents because of the possibility of core meltdown. As the reactor coolant drains out of the RCS, the temperatures of the nuclear fuel rods increase due to the lack of coolant. Core meltdown may result from the increased temperatures of the fuel rods.

An ECCS is one of the engineered safety features and supplies sufficient coolants to a core for maintaining fuel temperatures below its melting point and therefore core meltdown could be avoided in case of a LOCA. An ECCS consists of a safety injection system (SIS) and a shutdown cooling system (SCS). The purpose of the SIS is core heat removal and power decrease via borated water injection following a LOCA. In the APR1400, emergency coolant is injected from safety injection tanks (SITs) and via safety injection pumps (SIPs) from an in-containment refueling water storage tank (IRWST). The SCS is designed to provide residual heat removal in shutdown situations, which is a long term operation mode [12]

2.2 Nanofluids

Nanofluids are engineered colloidal dispersions with a traditional coolant as a base in which nanoparticles are suspended. In 1995, Choi, who first named nanofluids, published the results of his theoretical research. Subsequent developments in nanofluid engineering have contributed to the rapid growth in nanotechnology and surface technologies over the last 10 years [2].

The colloidal suspensions have substantially shown intriguing thermal performances regarding four points: (1) increased thermal conductivity (approx. 150%), (2) increased single-phase heat transfer coefficient (approx. 60%), (3) increased critical heat flux with extended nucleate boiling regime (approx. 200%), and (4) improved quenching efficiency. Although there is a lack of agreement of the experimental data in the literature and a lack of understanding of the physical mechanisms describing nanofluid thermal performances, the present work was motivated by the fact that a nanofluid formulation could not be tailored to show the desired properties for nuclear systems if we do not consider it together with nuclear safety systems characterized inherently by various couplings in a system engineering [13]. These properties were expected to be better especially when nanofluids are employed as coolants in ECCSs, and several applications to light water reactors have been published [4–7]. The design alternatives suggested in previous studies can be summarized as follows:

OPTION 1: injection of nanofluids from conventional SITs

OPTION 2: installation of a nanofluids-engineered SIT which is dedicated for nanoparticle injection

OPTION 3: injection of nanofluids via IRWST lines

Axiomatic design and TRIZ

The purpose of AD is to create the improved design of various tangible and intangible products. It provides an objective means for evaluating competing designs and enabling a better design to be chosen. The following two axioms are expressions that integrate common principles mapping FRs and DPs [9]:

Axiom 1: Independence Axiom. Maintain the independence of the FRs.

Axiom 2: Information Axiom. Minimize the information content of the design.

The mapping process between two domains can also be represented mathematically in terms of characteristic vectors that define the design goals or FRs and design solutions or DPs:

$$\{FR\} = [A]\{DP\}$$

where [A] is the design matrix.

In order to satisfy axiom 1, [A] must be either diagonal or triangular. In the case where [A] is diagonal, each of the FRs can be satisfied independently by one of the DPs. This situation is called an uncoupled design. In the case where [A] is triangular, the independence of the FRs can be satisfied by determining the DPs in the proper sequence. This is called a decoupled design. In this case, [A] does not necessarily have to be triangular in strict sense. A design corresponding to any other form of the design matrix is said to be coupled. Consequently, a diagonal or triangular design matrix gives rise to a better design. In practice, however, most designs are coupled, so that designers must decouple the object design by intelligently selecting relevant DPs, which can be referred as an improved design. Finally, axiom 2 can be used for determining the best solution among the decoupled or uncoupled design options.

Following the AD process, designers may encounter a problem about how to obtain a decoupled solution or convert a coupled design into decoupled. TRIZ is one of the tools to give a reasonable approach to find such a solution. TRIZ is a knowledge-based methodology for facilitating problem-solving in the context of invention. Specifically, by utilizing problem-solving elements from prominent inventions, one can solve one's own problems or produce further innovations with less effort. TRIZ is based on a hypothesis that patterns or analogical concepts leading to exceptionally innovative solutions to technical problems can be extracted through analysis of past inventions. In support of the hypothesis, an enormous number of patents have been researched and still being analyzed [10].

The most significant benefit of using TRIZ in AD process is identification and consequently removal of technical contradiction (which is the interchangeable term as coupling in AD process) with the help of certain principles. It should be noted that decoupling a design in terms of AD is similar to removing a technical contradiction in TRIZ. In a design process, the principles of AD do not practically provide a guaranteed method to determine the proper DPs. However, all of the TRIZ principles focus on methods to exclude technical contradictions within not only all branches of engineering, but also non-technical fields as well. Figure 1 shows how to apply TRIZ in AD to a problem.

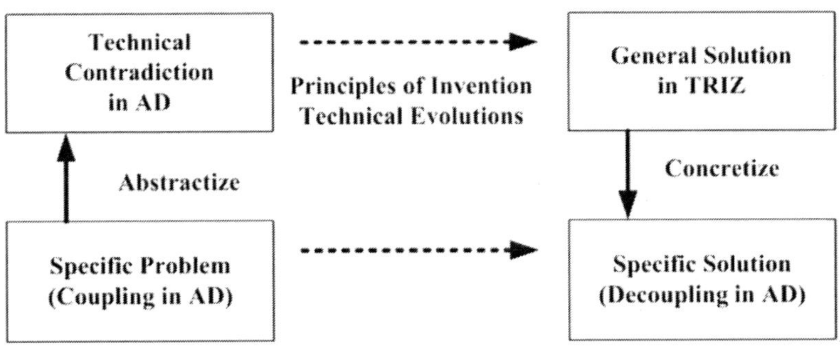

Figure 1 Problem-solving using TRIZ in AD process.

The coupling recognized in AD is equivalently associated with a contradiction with TRIZ. To create the DPs to decouple the design matrix, designers should represent their specific problems with typical language. This process, called 'abstractization', is the first step for applying the TRIZ on the results of AD process. Utilizing the TRIZ principles for abstractized problems or contradictions helps designers obtain general solutions which are the ideas to remove the technical contradictions by changing the physical parameters such as geometrical shapes, physical conditions, matter states, and so on. Since the general solutions are not directly applicable to the specific problems, mostly designers are required to concretize the general solutions for the specific solutions concerning the characteristics of the original problems [14,15].

2.3 Design process and results

The design process was started from the reverse engineering of conventional ECCS and nanofluids injection mechanisms using the principles of AD followed by identification of design weaknesses in terms of couplings. These design weaknesses was then eliminated by virtue of TRIZ, and finally a specific design was created.

2.4 Reverse engineering of conventional ECCSs

We investigated the FRs of the ECCS of the APR1400 to employ a nanofluid injection mechanism. Three nanofluid injection mechanisms were referred from previous studies [5–7]. The FR and DP decomposition starts at the top requirement of the ECCS as follows [14, 15]:

FR0: shut down a reactor while preventing core melt after LOCA

DP0: nanofluid-engineered ECCS

The sub-FRs of DP0 are the requirements to inject coolant during the initial phase (FR1) and to provide long-term cooling (FR2). The respective DPs of FR1 and FR2 constitute the SIS and SCS as follows:

FR1: injection of coolant at the beginning of LOCA

DP1: safety injection system

FR2: provides long-term cooling for reactor cold shutdown

DP2: shutdown cooling system

FR1 should be related to the reflooding and refilling stage. The corresponding DP1 is primarily associated with all options. With regard to option 1, FR1.1, FR1.2, and FR1.3 are as follows:

FR1.1: provides coolant from conventional SITs

DP1.1: injection of borated water from conventional SITs

FR1.2: provides nanoparticles

DP1.2: [Option 1] injection of nanofluids from conventional SITs as nanoparticles mixed with borated water

FR1.3: provides coolant from a conventional IRWST

DP1.3: injection of borated water from conventional IRWST

We can define the sub-requirements of DP1.1 again as follows.

FR1.1.1: provides sufficient coolant inventory

DP1.1.1: conventional SITs

FR1.1.2: provides a driving force for coolant

DP1.1.2: pressurized nitrogen

FR1.1.3: provides a control signal for coolant

DP1.1.3: passive actuation

FR1.1.4: provides a flow path for coolant

DP1.1.4: valve arrangement

A sub-FR of DP1.2, 'injection of nanofluids from conventional SITs as nanoparticles mixed with borated water' is required to provide a sufficient quantity of nanoparticles, stability of the nanoparticles, convenient nanoparticle sampling, compatibility with tanks, structures for nanoparticle storage, and continuous mixing with borated water. They may be summarized is as follows:

FR1.2.1: provides a sufficient quantity of nanoparticles

DP1.2.1: volume percentages of nanoparticles

FR1.2.2: provides continuous stability of nanoparticles

DP1.2.2: pH stabilizer (acidic condition)

FR1.2.3: provides a check for nanoparticle stability

DP1.2.3: sampling device

FR1.2.4: provides compatibility for corrosion resistance

DP1.2.4: neutral pH stabilizer

FR1.2.5: provides storage for nanoparticles

DP1.2.5: conventional SIT

FR1.2.6: mixes the nanoparticles and coolant

DP1.2.6: preloaded nanoparticles into a conventional SIT

We completed the set of FRs and DPs and Figure 2 shows the entire design matrix for option 1. The boxes marked in Figure 2 indicate that the functional couplings or technical contradictions. There are two important couplings identified due to the share of the conventional SIT as a reservoir of coolant and nanofluid:

$$\begin{bmatrix} FR1.1.2 \\ FR1.2.3 \end{bmatrix} = \begin{bmatrix} X & X \\ X & X \end{bmatrix} \begin{bmatrix} DP1.1.2 \\ DP1.2.3 \end{bmatrix}$$

where FR1.1.2 provides a driving force for the nanoparticles, FR1.2.3 provides a check for nanoparticle stability, andDP1.1.2 pressurized nitrogen, DP1.2.3 sampling device.

This is the first contradiction caused by nitrogen pre-pressurization and nanofluid sampling. The sampling for checking the stability of nanoparticles requires depressurization of the contents of the SIT. However, the depressurization of the SIT does not contribute to a driving force for the coolant and nanoparticles.

$$\begin{bmatrix} FR1.2.2 \\ FR1.2.4 \end{bmatrix} = \begin{bmatrix} X & X \\ X & X \end{bmatrix} \begin{bmatrix} DP1.2.2 \\ DP1.2.4 \end{bmatrix}$$

where FR1.2.2 provides stability of nanoparticles, FR1.2.4 provides compatibility for corrosion resistance, DP1.2.2 pH stabilizer (acidic condition), and DP1.2.4 neutral pH stabilizer.

The original option 1 does not mention any methods to control corrosion caused by nanofluids. To ensure homogeneity of the nanoparticles in the SIT, the pH state of the SIT should be acidic. However, acidic contents can induce corrosion of the SIT. Consequently, the FRs and DPs which satisfy each of these functions contradict one another.

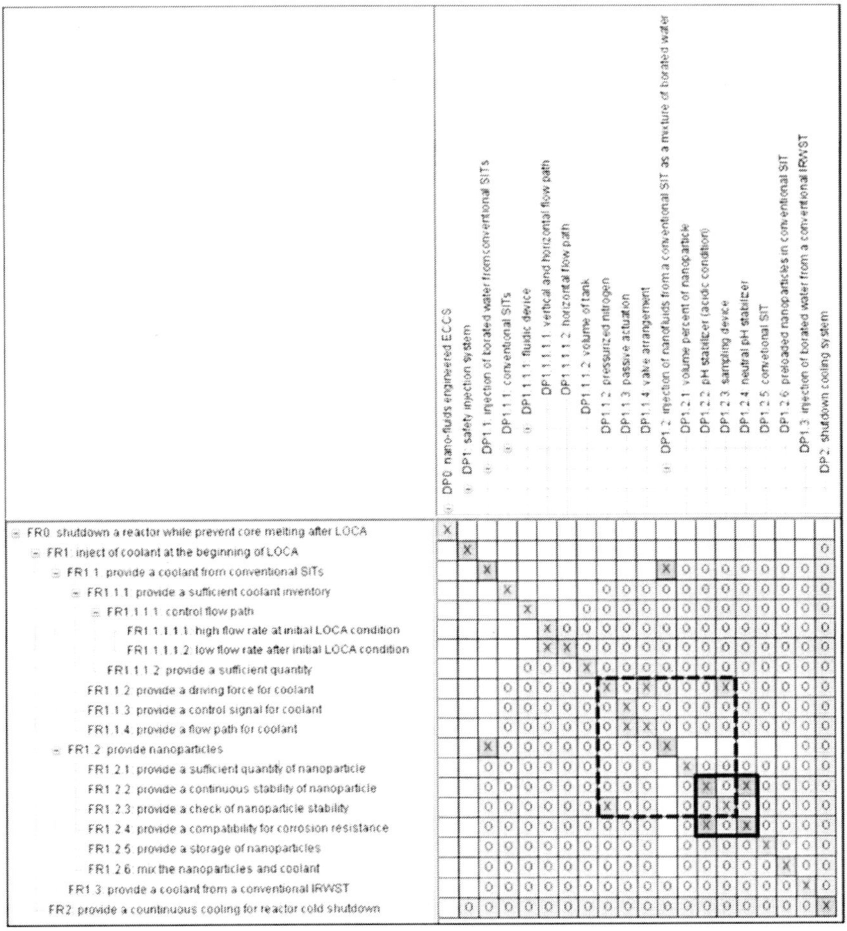

Figure 2. Design matrix for option 1 (dash: coupling 1, solid: coupling 2).

Using the same method, option 2 was then analyzed. Identifying the FRs and DPs for option 2 is similar to the process for option 1 except the facts that option 2 uses a separated nanofluid injection tank and the tank has titanium coating due to the corrosion of the tank wall. Since the size of the nanofluid injection tank is much smaller than the SITs, the wall coating is reasonable in terms of cost.

FR1.1: provides borated water

DP1.1: conventional SITs

FR1.2: provides nanoparticles

DP1.2: [Option 2] a nanofluid engineered SIT which is dedicated for nanoparticle injection

FR1.3: provides coolant from a conventional IRWST

DP1.3: injection of borated water from a conventional IRWST

Figure 3 shows the coupling in designing option 2. The coupling in option 2 is identical to the first coupling of option 1 since it also needs nitrogen pre-pressurization and nanofluid sampling.

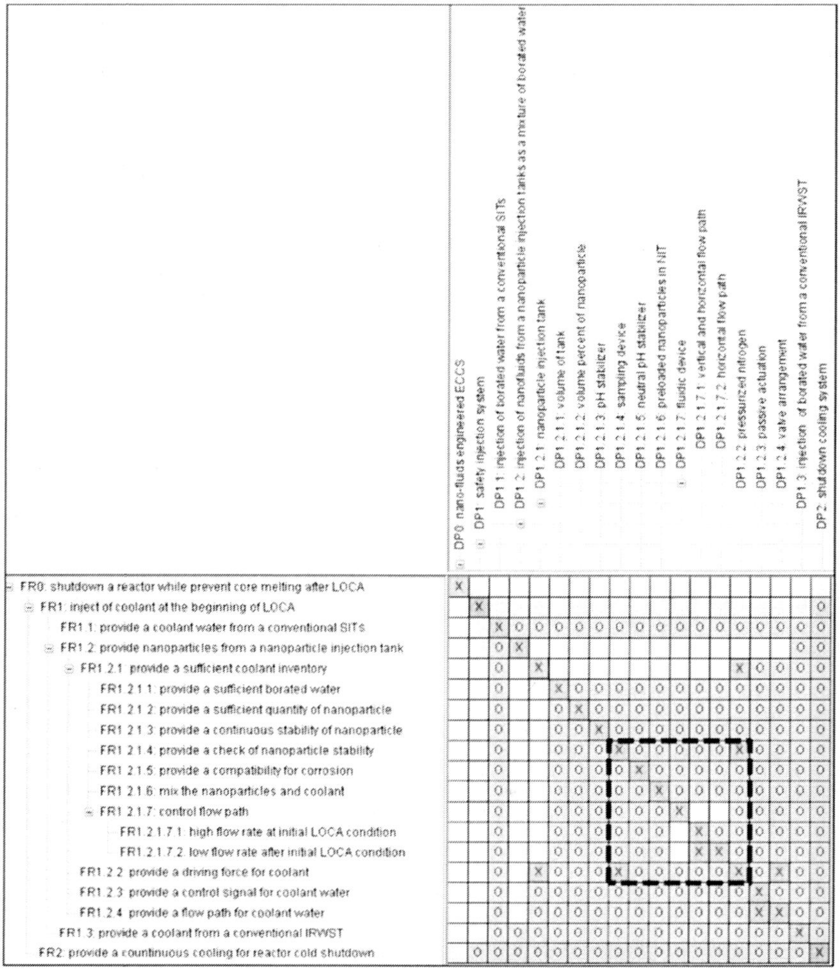

Figure 3. Design matrix for option 2.

Option 3 is the nanoparticle injection method connected with the IRWST. It provides the nanoparticles in a separate tank with coolant from the IRWST in a mixed state by the SIP. Figure 4 shows the resulting design matrix for option 3.

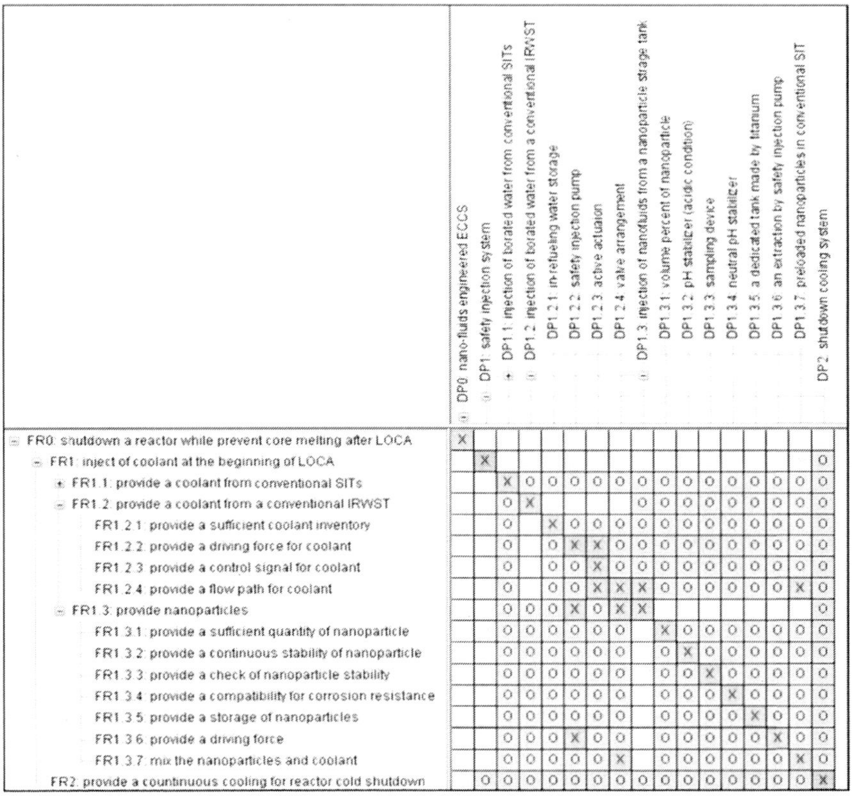

	DP0: nano-fluids engineered ECCS	DP1: safety injection system	DP1.1: injection of borated water from conventional SITs	DP1.2: injection of borated water from a conventional IRWST	DP1.2.1: in-refueling water storage	DP1.2.2: safety injection pump	DP1.2.3: active actuaion	DP1.2.4: valve arrangement	DP1.3: injection of nanofluids from a nanoparticle strage tank	DP1.3.1: volume percent of nanoparticle	DP1.3.2: pH stabilizer (acidic condition)	DP1.3.3: sampling device	DP1.3.4: neutral pH stabilizer	DP1.3.5: a dedicated tank made by titanium	DP1.3.6: an extraction by safety injection pump	DP1.3.7: preloaded nanoparticles in conventional SIT	DP2: shutdown cooling system
FR0: shutdown a reactor while prevent core melting after LOCA	X																
FR1: inject of coolant at the beginning of LOCA		X															O
FR1.1: provide a coolant from conventional SITs			X	O	O	O	O	O	O	O	O	O	O	O	O	O	O
FR1.2: provide a coolant from a conventional IRWST			O	X					O	O	O	O	O	O	O	O	O
FR1.2.1: provide a sufficient coolant inventory			O		X	O	O	O	O	O	O	O	O	O	O	O	O
FR1.2.2: provide a driving force for coolant			O		O	X	X	O	O	O	O	O	O	O	O	O	O
FR1.2.3: provide a control signal for coolant			O		O	O	X	O	O	O	O	O	O	O	O	O	O
FR1.2.4: provide a flow path for coolant			O		O	O	X	X	X	O	O	O	O	O	O	X	O
FR1.3: provide nanoparticles			O	O	O	X	O	X	X								O
FR1.3.1: provide a sufficient quantity of nanoparticle			O	O	O	O	O	O		X	O	O	O	O	O	O	O
FR1.3.2: provide a continuous stability of nanoparticle			O	O	O	O	O	O		O	X	O	O	O	O	O	O
FR1.3.3: provide a check of nanoparticle stability			O	O	O	O	O	O		O	O	X	O	O	O	O	O
FR1.3.4: provide a compatibility for corrosion resistance			O	O	O	O	O	O		O	O	O	X	O	O	O	O
FR1.3.5: provide a storage of nanoparticles			O	O	O	O	O	O		O	O	O	O	X	O	O	O
FR1.3.6: provide a driving force			O	O	O	X	O	O		O	O	O	O	O	X	O	O
FR1.3.7: mix the nanoparticles and coolant			O	O	O	O	O	X		O	O	O	O	O	O	X	O
FR2: provide a countinuous cooling for reactor cold shutdown		O	O	O	O	O	O	O	O	O	O	O	O	O	O	O	X

Figure 4. Design matrix for option 3.

In option 3, any significant couplings do not exist due to a separate reservoir for nanofluid and an active driving force such as SIPs.

2.5 Considerations for constraints

In order to finalize a design process, we should consider design constraints, for example, injection time, reliability, installation space, and cost for nanofluid-engineered ECCSs. The constraints belong to a detailed design process so it is not easy to estimate them quantitatively at this moment. We discuss the overall aspects of constraints on the basis of the analyzed FRs.

First, nanoparticles should be promptly injected into the core. From this viewpoint, options 1 and 2 are preferable because the SITs can be activated at an early stage of LOCAs. Option 3 may contribute to coolability at later stage. On the other hand, options 1 and 2 do not further provide nanofluids after the refill stage. Option 3 can provide a continuous nanofluid via a flow rate control device connected to the IRWST. Second, nanoparticles must be provided reliably in any types of injection mechanism. Option 1 can be regarded as the best from the viewpoint of risk due to its passive characteristic. Option 2 and 3 may need additional devices to isolate and pass the nanofluids in the tank in an appropriate manner, so reliability may be less than option 1. Third, option 1 does not require any additional space in the containment for installation as discussed. In practice, a large space may not be required for installation of a nanoparticle injection tank for option 2 and 3 because the volume of nanofluids is much smaller than that of emergency coolant. From the viewpoint of cost, installation space, additional manufacturing, maintenance may be associated. Since option 2 and 3 equips separate devices, it is likely to need higher cost.

2.6 Design modification

In this section, we reformed conventional nanoparticle injection mechanisms using TRIZ. First, we focused on the coupling discussed for option 1 and 2. For the first coupling, consisting of the FRs 'provide a driving force for the nanoparticles' and 'provide a check of nanoparticles stability' and the DP 'pressurized nitrogen' and 'sampling device', we abstractized the technical contradiction as 'pre-pressurization, but depressurization'. Then, we applied the principles of TRIZ to find the general solution of the technical contradiction with a contradiction matrix system. This matrix provides selective options for 'Features to improve' and 'Undesired results'. We applied our technical contradiction to this matrix by the feature to improve, 'the difficulty of detection and measurement' and the undesired result, 'the Shape'. The result of this matrix is summarized in Table 1.

We took the third and fourth principles in this case, though all principles in the results can be allowable. Considering the third principles, the conventional SITs and nanofluid injection tank should be separated. Then, by the fourth principle, SITs should be pressurized and the nanofluid injection tank should keep depressurized but can pressurize depending on conditions. At this time, the second coupling in option 1 was simultaneously resolved by adopting a titanium coating tank for nanofluids, which is the same method for option 2.

Table 1. Inventive principles in terms of the first coupling

Inventive principles	Description of inventive principles
1. Cheap short-living	Replace an expensive object with a multitude of inexpensive objects, compromising certain qualities
2. The other way around	Invert the action used to solve a problem. For example, instead of cooling an object, heat it. Make the movable parts fixed, and the fixed parts movable Turn the object 'upside down'
3. Segmentation	Subdivide an object into parts Make an object easy to disassemble Increase the degree of fragmentation of an object
4. Inert atmosphere	Replace a normal environment with an inert one Add neutral parts, or inert additives to an object

Finally we concretized a nanofluid-engineered ECCS with a separate nanofluid injection tank as shown in Figure 5.

The developed nanofluid-engineered ECCS consists of a borated water storage and a nanofluid storage. The borated water storage tank, which is the conventional SIT, can sustain a pressure of over 40 bars. On the other hand, the nanofluid storage tank can maintain its contents at atmospheric pressure, while also maintaining them at a stable low pH with a titanium coating on the inner surface of the tank, which is required to tackle the nanoparticle stability problem.

As shown in Figure 5, during normal operation, the borated water and nanofluids are separated by an isolation stopper connected to a floater inside the SIT. This isolation stopper is placed at the inner surface of the SIT by pressurized nitrogen and the floater with sealing the pipe line connecting the SIT and nanofluids injection tank. Following a LOCA, the pressure inside the RCS decreases below the pressure of the SITs, at which point emergency coolant is injected into the RCS. At this moment, the gravitational force generated by the floater opens the isolation stopper. Finally, the pressurized nitrogen forces the nanofluids, mixed with coolant from the SITs, to be injected. The decoupled matrix resulting from this design is as follows:

$$\begin{bmatrix} FR1.1.2 \\ FR1.2.3 \end{bmatrix} = \begin{bmatrix} X & O \\ O & X \end{bmatrix} \begin{bmatrix} DP1.1.2 \\ DP1.2.3 \end{bmatrix}$$

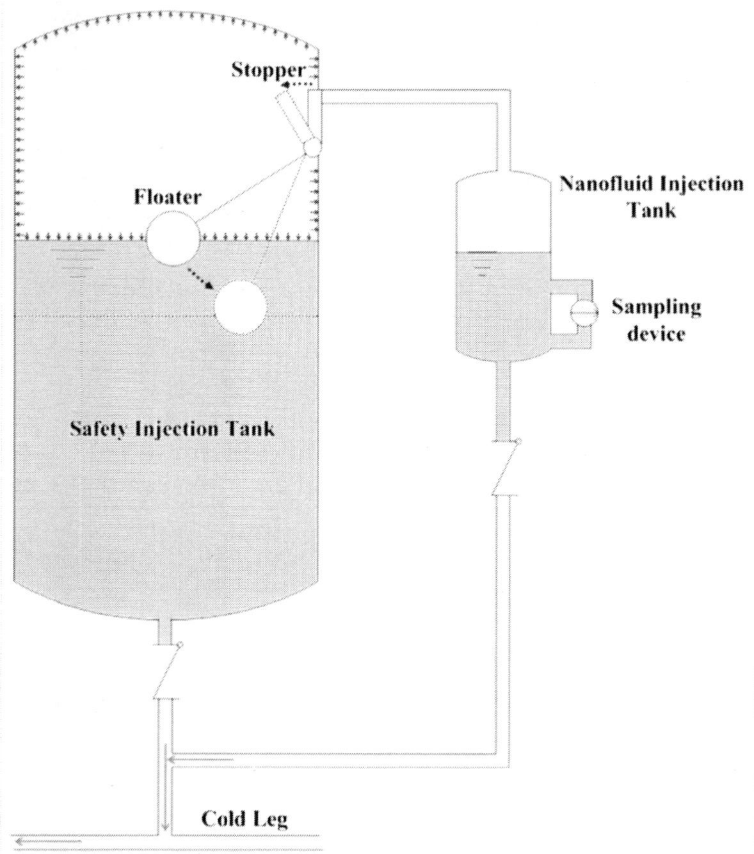

Figure 5. Schematic diagram of nanofluid-engineered SIT (out of scale).

where FR1.1.2 provides a driving force for the nanoparticles, FR1.2.3 provides a check for nanoparticle stability, DP1.1.2 pressurized nitrogen, and DP1.2.3 sampling device.

$$\begin{bmatrix} FR1.2.2 \\ FR1.2.4 \end{bmatrix} = \begin{bmatrix} X & O \\ O & X \end{bmatrix} \begin{bmatrix} DP1.2.2 \\ DP1.2.4 \end{bmatrix}$$

where FR1.2.2 provides stability to the nanoparticles, FR1.2.4 provides compatibility for corrosion resistance, DP1.2.2 pH stabilizer (acidic condition), and DP1.2.4 coating of anti-corrosion material in nanoparticle storage tank.

From the design matrix, the sampling device of the new system is no longer disturbed by pressurized nitrogen. Likewise, acidic condition of coolant will not harm to the tank anymore.

3. CONCLUSIONS

The history of nuclear industries should be the same as the history of nuclear safety. All activities from design to maintenance of NPPs are associated with safety. Even though the benefit of nanofluids is apparent in terms of heat transfer capability, its application should be considerate. It should be noted that both experience and theory have shown that the conceptual design stage is the most important for system's performance as well as safety, particularly in complex systems. Therefore, this study was performed to motivate and accelerate the use of nanofluid technologies in the practical applications.

The new method of nanofluids injection under LOCAs was proposed in this paper with analyzing the FRs of the candidates suggested in the previous researches on the basis of the principles of AD and TRIZ tools. Following the analysis of conventional ideas, major couplings were recognized. To solve these kinds of couplings, the installation of a separate nanofluids injection tank adjacent to the SITs was proposed. This tank is connected to the SIT with a passive gravity-operated stopper. The interior of the nanofluids injection tank is titanium-coated to permit a low pH environment. Depending on the detailed design processes, some of DPs may be revised or replaced, but this paper will contributed on the development of the preliminary steps connecting a scientific phase and an engineering phase.

ACKNOWLEDGEMENTS

This research was supported by Basic Science Research Program through the National Research Foundation of Korea (NRF) funded by the Ministry of Education, Science and Technology (2009-0069122).

COMPETING INTERESTS

The authors declare that they have no competing interests.

AUTHORS' CONTRIBUTIONS

MK participated in the reverse engineering and drafted the manuscript. CJ performed the design modification and participated in the reverse engineering. SP performed the precedent analysis and participated in the reverse engineering. IB provided the properties of nanofluids. GH conceived of the study, and participated in its design and coordination. All authors read and approved the final manuscript.

REFERENCES

1. Choi SUS: Enhancing thermal conductivity of fluids with nanoparticles. In *Developments and Applications of Non-Newtonian Flows*. New York: ASME; 1995:99–105. *FED-Vol.23*. FED-Vol.23.

2. Kim HD, Dewitt G, Mckrell T, Buongiorno J, Hu L: On the quenching of steel and zircaloy spheres in water-based nanofluids with alumina, silica and diamond nanoparticles. *Int J Multiphase Flow* 2009, 35: 427–438.

3. Kim SJ: Pool boiling heat transfer characteristics of nanofluids. *PhD thesis*. Massachusetts Institute of Technology; 2007.

4. Chupin A, Hu LW, Buongiorno J: Applications of nanofluids to enhance LWR accidents management in in-vessel retention and emergency core cooling systems. In *Proceedings of International Congress on Advances in Nuclear Power Plants 2008*. Anaheim, California; 2008.

5. Pop MGM, Brian GL: Nuclear power plant using nanoparticles in emergency systems and related method. 2008. U.S. Patent: US 2008/ 0212733A1

6. Pop MGM, Brian GL, Angelo B: Nuclear power plant using nanoparticles in emergency situations and related method. 2008. U.S. Patent: US 2008/0219395A1

7. Pop MGM, Brian GL: Nuclear power plant using nanoparticles in closed circuit of energency systems and related method. 2008. U.S. Patent US 2008/029396A1

8. Suh NP, Bell AC, Gossard DC: On an axiomatic approach to manufacturing, manufacturing systems. *J Eng Ind*1978, 100: 127–130.

9. Suh NP: *Axiomatic design: Advances and Applications*. Oxford: Oxford University Press; 2010.

10. Alshuller G, Shulyak L, Fedoseev U: *40 Principles TRIZ Key to Technical Innovation*. Worcester: Technical Innovation Center Inc; 1998.

11. Shin GS, Park GJ: Conceptual design of a beam splitter for the laser marker using axiomatic design and TRIZ. *J Korean Soc Precis Eng* 2004, 21: 166–173.

12. Lee SS, Kim SH, Suh KY: The design features of the advanced power reactor 1400. *Nucl Eng Technol* 2009, 41: 995–1004.

13. Bang IC, Heo G, Jeong YH, Heo S: An axiomatic design approach of nanofluid-engineered nuclear safety features for generation III+ reactors. *Nucl Eng Technol* 2009, 41: 1157–1170.

14. Heo G, Lee SK: Design evaluation of emergency core cooling systems using axiomatic design. *Nucl Eng Des*2007, 237: 38–46.

15. Ahmed R, Koo JM, Jeong YH, Heo G: Design of safety-critical systems using the complementarities of success and failure domains with a case study. *Reliab Eng Syst Saf* 2011, 96: 201–209

Chapter 3

LWR Safety Analysis and Licensingand Implications for Advanced Reactors

P. F. Frutuoso e Melo1[1], I. M. S. Oliveira1[1] and P. L. Saldanha2[1]

[1] 1COPPE/UFRJ – Programa de Engenharia Nuclear,, Brazil

1. INTRODUCTION

Most reactors under operation nowadays are light water reactors (LWR). The licensing and safety basis for them has been mainly deterministic. This approach has been under use since the beginning of commercial nuclear power in the 1950s. The purpose of this chapter is to discuss what this deterministic basis is, and how it has been used with emphasis on the US and German experience. This emphasis is because the first Brazilian reactor is of Westinghouse design, while the second one is of KWU (Kraftwerk Union)/Siemens/Areva design. Both designs are pressurized water reactors (PWR).

This chapter starts with the discussion of safety criteria, consideration of the defense in depth approach and deterministic criteria (safety margins), and the discussion of design basis accidents, including plant safety systems for meeting safety design criteria (IAEA, 2009a), Ahn et al (2010).

The approaches used thus far for safety analyses of LWRs have been essentially deterministic, where engineering judgment and conservatism have been used to face uncertainties. An example of this approach is the consideration of design basis accidents (such as large loss of coolant accidents – LOCA). They have been defined by arbitrarily combining initiating events with single failures (for example, loss of an injection pump), Kim et al (2010).

The inception of risk-informed decision making was in the 1970s, with the publication of the Reactor Safety Study NRC (1975), although it was initially named probabilistic risk assessment in the US. Since then, many improvements have been achieved but risk criteria have not as yet been established.

The risk-informed approach has been adopted by the US Nuclear Regulatory Commission (NRC) as an aid in the licensing and safety basis of US nuclear power plants. This means that the formal licensing process is to be approached by deterministic and probabilistic methods. The risk-informed approach may represent the formal presentation of a level 3 probabilistic safety assessment (PSA), so that plant risk curves are available. However, regulators do not as yet have risk criteria for this purpose, so that PSAs are recommended but their results are not compared to any criteria. Instead, there is a criterion concerning level 1 PSAs (in this case, the reactor core degradation frequency is estimated), and this is the central feature of risk-informed decision making nowadays, NRC (2002).

The risk-informed approach will be discussed in the light of US plants experience. Approaches in other countries will also be presented and the gained experience will be commented, Kadak & Matsuo (2007).

2. BASIC DEFINITIONS

Safety analysis is the study, examination and description of a nuclear installation expected behavior throughout its life, under normal and transient conditions, and also under postulated events, in order to determine: a) safety margins provided in normal operation and in transient regimes; b) the adequacy of items to prevent the consequences of accidents that may occur.

Safety assessment is the systematic and independent evaluation carried out by the regulator on the submitted safety analysis. It is used to support and subsidize the licensing decision on the plant acceptability, once the risk associated with its operation is known.

Inspection is the core activity performed by the regulator to verify compliance with its regulatory requirements or expressed in terms of licenses and permits and to implement them through coercive actions.

Regulatory inspection is the examination, observation, measuring, testing and verification of documentation executed by the regulator during any stage of the licensing process, to ensure compliance of materials, components, systems, structures, operational activities, processes, procedures and qualification of personnel with pre-established requirements or determined by the regulatory body.

Figure 1 displays the general Brazilian licensing process, where the definitions discussed herein play an important role.

The Preliminary Safety Analysis Report (PSAR) is one of the reports of the application for a building permit. It aims at demonstrating that the applicant is qualified to manage the construction application by providing design-related technical information to the regulator.

The Final Safety Analysis Report (FSAR) presents the final analysis and evaluation of plant design, as constructed, and the behavior of items in order to assess the risk to health and safety of the population as a whole, resulting from the installation and considering the information provided since the presentation of the PSAR.

A discussion on PSA meaning will be presented in Section 7 in the context of risk-informed decision making.

3. DEFENSE IN DEPTH

The basic philosophy of nuclear power plant design has been described as defense in depth expressed in terms of three safety levels (NRC). These levels cover a variety of considerations that often intertwine, so that the allocation of certain aspects of the project to one level or another is somewhat arbitrary. However, these levels are useful to indicate the various stages in the safety design of a nuclear power plant. NRC defines each safety level through ordinances or rules. The basic purpose of reactor safety is to maintain the integrity of multiple barriers against the release of fission products. This integrity is supported by a defense in depth approach on three safety levels: a) prevention; b) protection; c) mitigation.

However, it has been more convenient to subdivide the second safety level (protection) into three new levels. Therefore, the implementation of the defense in depth concept is made through five levels whose objectives and essential means are displayed in Table 1, IAEA (1996). The role of each safety level can be clearly seen in the table. One desirable effect of the defense in depth concept is that the plant that adopts it tends to be more resilient to failures.

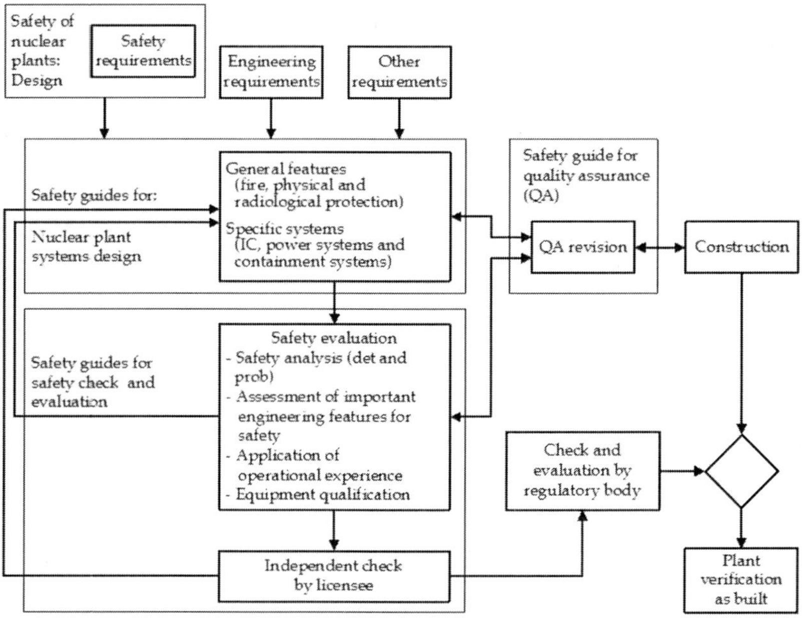

Figure 1. General Brazilian licensing process

Table 1. Objectives and essential means of the defense in depth approach, IAEA (1996)

Level	Objective	Essential means
1	Prevention of abnormal operation and/or failures	Conservative design High quality in construction & operation
2	Control of abnormal operation and detection of failures (protection)	Control systems Limiting systems Protection systems
3	Accident control within design basis (protection)	Engineered safety features Accident procedures
4	Control of severe plant conditions (protection)	Complementary measures Accident management
5	Mitigation of radiological consequences of significant radioactive releases	Off-site emergency response

4. ACCIDENT ANALYSIS

The construction and operation of nuclear power plants requires the submission of a safety analysis report which must contain an analysis of a wide range of conceivable abnormal events. The purpose is to demonstrate that the project provides a means to control these events or otherwise accommodate their consequences without undue risk to health and safety of the public.

Analyzed conditions include: a) small transients that occur with moderate frequency and represent minor hazards; b) unlikely accident situations that can have serious consequences and therefore require different measures to protect the public.

Safety analysis is concerned with the potential effects of every conceivable (or anticipated) transient that may occur as a result of: a) operational malfunctions, e.g., human errors or small instrumentation or other equipment failures, or b) serious mechanical failures of different types.

Transients of moderate frequency can result from operational occurrences (or other), which create an imbalance between heat generation in the fuel and its removal: *a*) thermal power increase, caused by:*a*.1) decrease of coolant temperature; or *a*.2) removal of control material (burnable poisons); *b*) decrease in cooling efficiency.

As to low frequency events, there can be: a) small pipe ruptures; b) loss of flow accidents (LOFA); and c) design basis accidents (DBA).

Small pipe ruptures are more serious when they occur in an input line of the pressure vessel of a PWR primary system circuit. The reactor is shut down by the reactor protection system (RPS) but there is loss of water to the containment (vapor flashing also occurs). In general, for breaches of equivalent diameter smaller than 0.5", the chemical and volume control systems (CVCS) compensates for inventory losses of the reactor cooling system (RCS).

Should a loss of off-site and on-site power occur, all pumps eventually stop and the result is a loss of flow accident (LOFA). However, in 10s, in general, power will be available through emergency diesel generators. Meanwhile, the reactor is shut down when receiving a loss of flow signal, and steam is removed automatically from the turbine (steam dump). As there is some energy production during steam withdrawal, recirculation pumps typically remain connected to the main generator bus for about 10 seconds. Recirculation during pump shutdown and some natural circulation of coolant is usually sufficient to prevent the condition of critical heat flux after reactor trip.

Design basis accidents involve the postulated failure of one or more major systems and an analysis based on conservative assumptions (e.g., pessimistic estimates of fission product releases). It must be shown that the radiological consequences are within preset limits. These accidents serve as a basis for assessing the general acceptability of a particular reactor design. Design basis accidents are classified as Knief (1993): a) overcooling - heat removal increasing on the secondary side; b) subcooling - reduced heat removal on the secondary side; c) overfilling - increased inventory of reactor coolant; d) loss of flow - RCS (reactor coolant system) descreased flow; e) coolant loss - loss of reactor coolant inventory; f) Reactivity - reactivity and power distribution anomalies in reactor core; g) ATWS - anticipated transients without scram; h) Spent fuel and waste system - radioactivity release from spent fuel element or a subsystem or reactor component; i) external events - natural or man-made events that can affect plant operation and safety systems.

A major break in a steam line results in a reactivity insertion of cold water (overcooling) systems in several loop systems. This event causes liquid flashing in the secondary side of steam generators. The secondary fluid cools by removing heat from the primary (overcooling), with important implications for the reactivity balance.

In accidents related to overcooling, or others that require rapid reduction in temperature in support of depressurization, the pressurized thermal shock (PTS) phenomenon is a concern of great importance. It is a boundary condition of reactor vessel integrity. It may occur during a system transient that primarily causes severe overcooling of the vessel wall inner surface and then results in high repressurization. If there is significant degradation due to radiation embrittlement and if there are defects of critical sizes in the vessel wall, this may fail. PTS is prevented by operating within boundary curves of temperature-pressure which are periodically revised to reflect the vessel current condition, particularly in terms of radiation embrittlement. This approach tends to lead to increasing restrictions on the operation window for plant heating (heatup) and cooling (cooldown) as the plant ages.

The anticipated transient without scram (ATWS) has two general characteristics: a) it starts through a transient whose occurrence is anticipated one or more times in reactor life; b) posterior reactor trip does not occur (that is, a failure occurs). This failure, especially a reactivity insertion (control rod removal) is solved by negative reactivity feedbacks that diminish the reactor power level, or at least diminish its growth. Adequate reliability of control rods and the reactor protection system are important to prevent such events.

A large rupture or leak in one or more steam generator (SG) tubes of a PWR results in a particular loss of coolant accident (LOCA) scenario because primary coolant passes directly to the secondary side. In addition to being radioactive, the coolant also represents an irretrievable loss of inventory in the containment building. The response to this accident includes isolation of damaged generators and rapid cooling and depressurization, to reduce the coolant loss, where care must be taken to avoid other accidents (e.g., PTS).

A loss of coolant accident (LOCA) occurs in general when there is loss of inventory in the primary system through a rupture of equivalent diameter larger than 0.5 "(for ruptures with equivalent diameter less than 0.5", the chemical and volume control systems (CVCS) compensates for inventory losses. Three types of LOCA are typically considered: a) small LOCAs: for equivalent rupture diameters between 0.5" and 3"; b) medium LOCAs: for equivalent rupture diameters between 3" and 6"; c) large LOCAs; for equivalent rupture diameters between 6" up to the double-ended or guillotine break in a reactor coolant system (RCS) cold leg, being this rupture considered as one of the design basis accidents.

The events that occur within the first 2 min following a design basis LOCA in a PWR are: a) blowdown: in which the reactor coolant is expelled from reactor vessel; b) refill: when emergency cooling water begins to fill the reactor vessel starting from the core bottom; c) reflood: when the water level raises enough to cool all reactor core.

In general, the emergency core cooling system (ECCS), one of the engineered safety features, should be designed to fit the following criteria under a postulated design basis LOCA in a PWR: a) the calculated maximum cladding temperature after the accident should not exceed 2200 °F (1204 °C); b) the calculated total cladding oxidation due to interaction of zircaloy with hot steam should not exceed 17% of the total cladding thickness before oxidation; c) the total amount of H_2 generated shall not exceed 1% of the hypothetical amount generated if all cladding material around pellets reacted; d) calculated changes in geometry, e.g., diameter of fuel rods and spacing should be such that the core can still be cooled; e) the calculated core temperature, after successful ECCS starting, must be maintained appropriately low for the time necessary for the decay of long half-life fission products in reactor core. More details on LOCA analysis may be found in Glasstone & Sesonske (1994).

Companies that sell reactors must provide analysis tools through which one can establish that the proposed reactor is designed to meet the criteria for emergency core cooling. These tools are generally complex computer programs that use thermal hydraulic models for calculating fuel and cladding temperatures, and other relevant situations and reactor characteristics. These tools should include means for calculating: a) energy sources; b) hydraulic parameters; c) heat transfer mechanisms of various hypothetical accident stages.

Different calculation programs have been developed and are being refined in order to calculate characteristic parameters, such as: a) coolant flow rates; b) enthalpy; c) coolant, fuel, and cladding temperatures; d) system pressure, under steady state and transient conditions.

Central to the above calculations is the notion of nodalization. Real reactor circuits must be nodalized, that is, a set of nodal volumes and junctions are defined and inserted into calculation programs to perform the desired safety calculations. An example of these nodalization procedures may be found inBorges et al (2001) concerning Angra 2 power plant.

5. SEVERE ACCIDENTS AND ACCIDENT MANAGEMENT

Severe accidents are those which are characterized by at least an initial core damage, typically specified as the overcoming of regulatory fuel limits, as, for example, 1200°C in the fuel cladding, as discussed in Section 4.

The need for considering severe accidents became apparent upon the issuance of the Reactor Safety Study (which will be briefly discussed in Section 7), NRC (1975), where a probability per year of the order of 1 in 20,000 reactor-years was estimated for core melt. This value was apparently higher than the one implicitly estimated for the reactors operating at that time (Petrangeli, 2009). This calculated figure meant an expected core melt each 40 years, although the Reactor Safety Study itself estimated that only one in about 100 core melt events could cause severe health consequences (up to 10 causalities). It is noteworthy that the Three Mile Island event reinforced and confirmed the need initially arisen for progress in nuclear safety by considering possible events beyond design basis.

IAEA (2000a) defines a severe accident as a very low probability plant state beyond design basis accident condition (like those discussed in Section 4), which may arise due to multiple failures of safety systems leading to significant core degradation. These failures may jeopardize the integrity of many or all of the barriers to the release of radioactive material.

IAEA (2000a) also mentions that the consideration of severe accidents shall not be performed as design basis accidents are, that is, by assuming conservative assumptions. Rather, realistic or best estimate assumptions, methods and analytical criteria should be employed.

In this sense, important event sequences that may lead to severe accidents shall be identified using a combination of probabilistic and deterministic methods and engineering judgement. Next, these event sequences are to be reviewed against a set of criteria aimed at determining which severe accidents shall be addressed in safety analysis.

Accident management has arisen to cope with severe accidents. IAEA (2000b) establishes some requirements on severe accident management and accident management in the operation of nuclear power plants. According to this, plant staff shall receive instructions in the management of accidents beyond design basis.

Examples of event sequences for PWRs in this context have been considered in the Reactor Safety Study (NRC, 1975), as a large-break LOCA with loss of all ac power and a transient-induced accident. This latter is caused by an event that requires reactor trip combined with a station blackout, i.e, the loss of all power, as well as the loss of capability of the secondary system to remove heat from the primary circuit.

External events might also play an important role in severe accident management since they are an importance source of energy for the reactor (Knief, 1992).

IAEA (2009b) discusses severe accident management programs for nuclear power plants. D'Auria & Galassi (2010) discuss important features on scaling in nuclear reactors that might be relevant for severe accident management. As mentioned earlier, as best estimates are to be used in severe accident management rather than conservative estimates, uncertainty analysis plays a dominant role in this field. Na et al (2004) present an approach for the prediction of major transient scenarios for severe accidents in nuclear power plants by using artificial intelligence.

6. LICENSING OF NUCLEAR POWER PLANTS

6.1. Introduction

The licensing of nuclear power reactors is a formal activity that constitutes a permanent process of decision making, involving the issuance of licenses, permits, amendments or their cancellations, covering issues involving the safety of nuclear reactors, and the radiological protection of operators, the general population and the environment.

Decision making is performed based on the results of two complementary activities: a) safety assessment; and b) inspection.

The decision should consider whether there is sufficient assurance that the facility operation will not result in undue risk to: a) population, b) operators and c) the environment.

The licensing process of nuclear facilities is regulated by standard CNEN-NE-1.04 (CNEN, 1984), in force since 1984. The issuance of licenses or permits shall be preceded by the applicant request together with information, data, plans and reports, whose content is described in the standard.

6.2. Applicable Standards

There are over 40 standards in force in CNEN (Brazilian Nuclear Energy Commission), and 20 apply to nuclear power reactors. In the absence of appropriate standardization, codes and guidelines of the International Atomic Energy Agency (IAEA), are preferably used, where necessary. Table 2 displays the most important nuclear standards concerning nuclear power reactors issued by CNEN. These standards may be found in cnen.gov.br.

6.3. The Licensing Process

The licensing process requires the issuance by CNEN of the following acts: a) Site Approval (AL); b) Construction License (LC); c) Authorization for Nuclear Material Use (AuMN); d) Authorization for Initial Operation; e) Authorization for Permanent Operation (AOP).

The various reports and programs per act required during the licensing process are presented below.

For site approval: a) Site Report; and b) Preliminary Program of Pre-Operational Monitoring.

For the Construction License (LC): a) Preliminary Safety Analysis Report (PSAR); b) Preliminary Plan of Physical Protection (PPPF); c) Quality Assurance Program (QAP); and d) Preliminary Plan for Personnel Training.

The following activities do not depend on a previous license: a) site excavation; b) infrastructure preparation; c) buildings not intended for safety-important items; and d) system components manufacturing.

Obligations during plant construction: a) report of deficiencies in the executive project, construction and pre-operational phase with impact on safety; b) progress report of activities; c) results of the programs of research and development (R & D) designed to solve safety problems; d) reports on equipment storage; e) audit programs on contractors; f) procedure for pre-operational tests, and g) submit to resident construction inspection.

Table 2. Typical CNEN standards for nuclear power reactors

Number	Title
NE-1.01	Reactor Operator Licensing
NE-1.04	Licensing of Nuclear Installations
NN-1.12	Qualification of Technical Independent Oversight Bodies in Nuclear Facilities
NE-1.14	Report of Nuclear Plants Operating
NN-1.15	Independent Technical Supervision in Quality Assurance Activities
NE-1.16	Quality Assurance for nuclear-power plants
NE-1.17	Personnel Qualification and Certification for Non-Destructive Testing Items in Nuclear Facilities
NE-1.22	Meteorological Programs in Support of nuclear-power plants
NE-1.26	Safety in Operation of nuclear-power plants
NE-2.01	Physical Protection of Nuclear Operating Units of Area
NN-2.03	Fire Protection in nuclear-power plants
NE-3.01	Basic Guidelines for Radiation Protection

Authorization for Initial Operation (AOI): a) Final Safety Analysis Report (FSAR); b) answers to LC constraints; c) authorization for nuclear material use; d) final plan for physical protection (FPF); e) radiation protection plan; f) fire protection plan; g) commissioning program; h) test procedures; i) Quality Assurance Program (PGQ); j) operating procedures manual; k) local emergency plan (PEL); l) operator team licensed by CNEN; m) civil responsibility insurance against damages; and n) submit to resident inspection.

Authorization for Permanent Operation (AOP): a) initial report of operations; b) commissioning report, and c) responses to AOI requirements.

During Operation: a) periodic reports; b) operational event reports; c) report to CNEN in Emergencies; d) shutdown planning; e) technical specification changing requests; f) technical modification requests; g) operator licenses reassessment; h) safety periodic review (each 10 years); i) response to CNEN requirements; j) submit to periodical inspections; and k) submit to resident inspection.

For safety review and assessment activities, four basic procedures are used: a) comparison with other facility used as a reference; b) verification of requirement, standard, and specification adherence; c) design verification through independent calculations; and d) incorporation of requirements arising from international experience in nuclear technology.

The verification of compliance requirements is made through a detailed examination of normative and support documents, identifying clearly the criteria that support the regulator assessment.

The analysis of the document or activity being evaluated is performed by comparing it with the regulator assessment criteria and/or previous requirements issued, following proper procedures for each type of task, such as: a) operational event; b) modification project; c) technical specification changes; d) Accident Analysis; e) periodical reports; and f) system and component design.

Next, a balance of deficiencies and nonconformities is performed.

The final product of the safety assessment is a technical advice. This document must contain the basis of judgement and conclude in a clear and concise way on the acceptability of the document or the activity under review. If there are deficiencies or nonconformities requirements for the implementation of corrective actions should be issued.

The objectives of independent calculations are: a) verify the completeness and adequacy of the analysis performed by the designer; and b) provide the regulator technical staff with experience and knowledge about phenomena and modeling techniques associated with the facility operation in normal or accident conditions.

Lessons learned through international operating experience and nuclear accidents are permanent sources of improvement of licensing requirements adopted by CNEN.

An inspection activity is made throughout all licensing phases, through testimonies, inspections and audits. Inspections may be reactive or routine. Reactive inspections (advised or not) are dependent on the project phase or on the occurrence of a significant event that requires verification. For reactors in permanent operation routine checks follow a regular program, which is established on an annual basis.

Regulatory Inspections are formal activities conducted by a team of inspectors which follows a previously prepared checklist, considering: a) inspection requirements (standards, license or permit terms, etc); b) examination of documents that regulate the inspected activity, such as: b.1) quality assurance program; b.2) operation manual; b.3) technical codes or standards; b.4) design specifications; b.5) FSAR applicable sections; b.6) checking of requirements not fulfilled in previous inspections.

During plant construction and operation phases, CNEN keeps a team of resident inspectors, which makes a plant daily monitoring and issues periodical audit reports. These reports describe inspection activities, identify non-compliances and formulate proper requirements for the licensed facility to deploy appropriate corrective actions, when necessary. Figure 2 display CNEN's inspection approach.

Tasks of power reactor licensing are performed through acts. These acts are related to the different steps during the licensing process: a) pre-licensing; b) site approval; c) construction issuance; d) during construction; e) AOP Issuance; f) operation monitoring. Acts related to pre-licensing involve: a) management contacts; b) verification of project objectives and preliminary schedules; and c) team meetings on licensing, quality systems and safety analysis.

Acts related to site approval involve: a) site report assessment (demographics, seismology, hydrology, meteorology, geography, and external events); b) emergency plan viability; and c) interaction with the environmental licensing (through the Brazilian environmental agency, IBAMA).

Acts related to construction issuance involve: a) PSAR examination and evaluation to check the safety concept acceptability of the plant design (design basis accidents, philosophy, design approach, experimental support, safety research, reference plant, standards adopted in the design and fabrication, program quality assurance and development of major providers, training program for human resources) ; and b) assessment of the pre-operational environmental monitoring program.

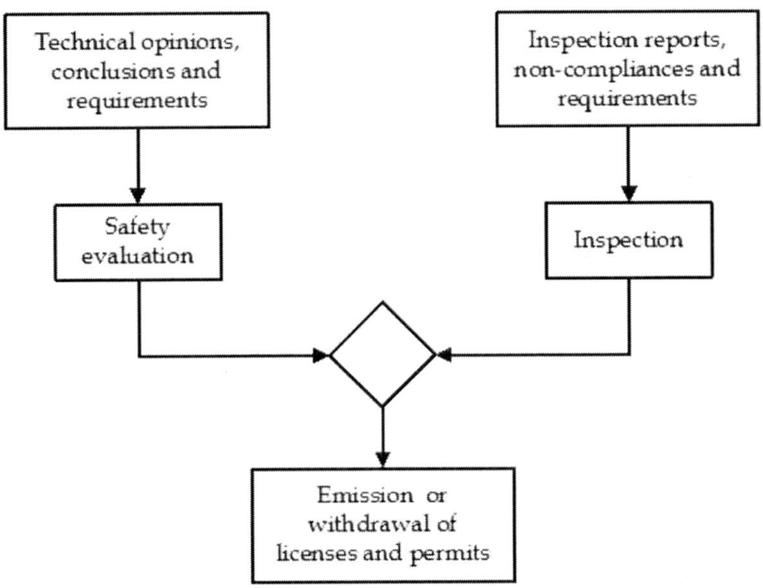

Figure 2. Brazilian nuclear regulator (CNEN)'s inspection approach

Acts during construction: a) assessment of safety deficiencies identified during the execute design, construction, assembly or pre-operational tests, from non-conformities recorded in the context of the Quality Assurance Program, or from deviations from the criteria and design basis as stated in PSAR, or arising from significant damage during construction, assembly or testing; b) FSAR review to check whether the design final specification confirms safety analysis findings; c) implementation inspection of procedures established in QAP, facility compliance as constructed in relation to licensed design, test adequacy on structure and system integrity as well as functional tests of components and systems; d) monitoring of international experience, with emphasis on the reference installation, to identify any additional measures that need to be required to improve safety of the facility under construction.

Acts during AOP issuance: a) assessment of compliance with all LC and AOI conditions; b) assessment of compliance with all CNEN safety significant requirements in earlier stages; c) beginning of resident inspection; d) procedure analysis and witness of integrated tests including loading tests; e) initial criticality; f) low power physical tests and other tests; g) initial operation report (ROI) evaluation to determine the adequacy of commissioning program to demonstrate foundations of safety analysis; h) survey of international safety standard and licensing evolution since the last license or permit issued.

Acts related to operation monitoring: a) resident inspection to verify compliance with terms set out in the AOP, particularly in relation to technical specifications; b) safety assessment on requirement and restriction compliance expressed in AOP; c) conduction of periodic inspection and audit program on activities that affect quality and are safety significant; d) assessment of operational safety by examining periodic operation reports, of consolidation of CNEN issued requirements and the examination of significant event reports; e) control and daily record of operational activities; f) assessment of technical change applications to be introduced in the licensed project or technical specifications changes; and g) monitoring of international operating nuclear reactors experience.

6.4. Psar and Fsar

The minimum content of PSAR comprises: *a*) Description and safety analysis of the site for the facility; *b*) Facility description and analysis with special attention to design features and operation; *c*) Preliminary design of the facility, with emphasis on: *c*.1) the main criteria;*c*.2) the design bases and their relationship with the main criteria, and *c*.3) information related to building materials, arrangement and approximate dimensions; *d*) Preliminary analysis and evaluation of project performance and installation of items in order to assess the risk to health and safety of people (safety margins for normal operation and transient conditions and adequacy of the items designed for accident prevention); *e*) Description and justification of the choice of variables based on the analysis and preliminary assessment that will be subject to technical specifications, and *f*) description of control systems for release of effluents and radioactive waste.

FSAR must include information that: *a*) describes the facility; *b*) provides the basis for the project; *c*) defines the limits of operation, and *d*) allows a safety analysis of the installation as a whole.

FSAR should allow for a: *a*) perfect understanding of the system design; and *b*) clear display of the relationships between the system design and safety assessments.

FSAR should also contain information relating to plant operation, like: *a*) quality assurance;*b*) program of pre-operational tests and initial operation; *c*) program for the conduct of operation, including: *c*.1) maintenance; *c*.2) periodic tests of items, and *d*) proposed technical specifications (TS).

Table 3 displays the FSAR contents.

Chapter 17 of FSAR is the only one written in Portuguese for Brazilian power plants, because all FSAR chapters except this one are prepared by the vendor. The chapter on quality assurance is prepared by the licensee itself.

A chapter 19 on probabilistic safety assessment (to assess core melt frequency, the so called Level 1 PSA as will be discussed in Section 7) is to be added to FSAR for Brazilian power plants.

6.5. Licensing of Angra 1 Nuclear Plant

Angra 1 has had its license covered by CNEN NE 1.04 and has been based on the American model of the Nuclear Regulatory Commission (NRC).

The operation time of 40 years was used in the project and considered in the safety assessment review for issuance of the Provisional Authorization of Operation (APO) in 1984, and later in the Authorization for Initial Operation (AOI) in 1987, and Authorization for Permanent Operation (AOP) in 1994.

In AOP, the time of 40 years was considered as a basis for 1984 and a review of the authorization to ratify or amend its terms is scheduled every 10 years. This ensures a periodical safety assessment review, keeping the licensing bases of CNEN–NE–1.26 standard.

Table 3. FSAR contents

Chapter	Contents
01	Introduction and General Description
02	Site Characteristics
03	Design of Structures, Components, Equipments & Systems
04	Reactor
05	Reactor Coolant Systems and Connected Systems
06	Engineered Safety Features
07	Instrumentation and Control
08	Electric Power
09	Auxiliary Systems
10	Steam and Power Conversion System
11	Radioactive Waste Management
12	Radiation Protection
13	Conduct of Operations
14	Initial Test Program
15	Accident Analysis
16	Technical Specifications
17	Garantia de Qualidade (Quality Assurance)
18	Human Factors Engineering

General Design Criteria adopted are described in Appendix A of 10 CFR 50, and were the minimum requirements for Angra 1 main criteria. The establishment of a defined accident spectrum that has been postulated for the project, whose consequences could not exceed the maximum dose limits on the borders of the "exclusion area", according to 10 CFR 100, characterized the deterministic licensing model.

The exclusion area is defined as the area in which an individual located at any point on its edge for 2 hours immediately after the release of fission products, would not receive a whole body radiation dose greater than 25 rem or a total thyroid radiation dose greater than 300 rem due to iodine exposure (Lamarsh & Baratta, 2000).

The verification of requirements established pursuant to 10 CFR 50 was driven by regulatory guides that consolidate the positions adopted and accepted by NRC technical assessment teams. FSAR standard model, as provided in standard NE-1.04, was the Regulatory Guide RG 1.70, Standard Format and Content of Safety Analysis Report for NPPs (1978). NUREG 0800, Standard Review Plan for Review Safety Analysis Report for NPP, is employed by CNEN for safety assessment.

6.6. Licensing of Angra 2 Nuclear Plant

Just as Angra 1's, Angra 2's licensing is subject to standards CNEN–NN-1.04 and 1.26. There is a direct correspondence between the American and German licensing models. To maintain uniformity between both Angra 1 and Angra 2 licensing, the FSAR contents, as provided in standard CNEN-NE-1.04 (CNEN, 1984) is in accordance with RG-1.70 (NRC, 1978), as amended to incorporate the developments in NUREG 0800 (NRC, 1996).

As will be discussed in Sec. 6.7, a noteworthy feature of Angra 2 licensing is the inclusion of human factors.

The safety criteria document presents the German Interior Ministry requirements that can be understood as minimum criteria in relation to the plant main design criteria. Guidelines for PWR reactors have recommendations for different design, divided into 25 chapters and, where applicable, they take into account technical standards from others, like ASME.

6.7. Human Factors and Human Reliability

A point worth mentioning is the incorporation into FSAR of the so called human factors engineering (Chapter 18). NUREG 0711 (NRC, 2004) has been adopted as a reference for the safety evaluations, taking into account the technological differences between Westinghouse and Siemens/KWU (AREVA) designs.

The human factors engineering approach to be presented in FSAR is composed by the following topics: a) Human factors engineering program management; b) Operating experience review; c) Functional Requirements Analysis and Function Allocations; d) Task Analysis; e) Personnel Qualification and Quantification; f) Human Reliability Analysis; g) Human – System Interface Design; h) Procedures Development; i) Development of the Training Programs; j) Human Factors Verification and Validation.

Figure 3 displays the NRC human factors engineering approach that has been adopted by CNEN.

6.8. Licensing in US

The Brazilian nuclear regulation was strongly influenced by the model used in the U.S., particularly with regard to stages of the licensing process. The basic law to regulate nuclear power is the Atomic Energy Act, 1954. In 1974, through the Energy Reorganization Act, an exclusive agency was created to regulate the use of nuclear energy, called the Nuclear Regulatory Commission (NRC).

The Code of Federal Regulations (CFR) is the collection of US technical documents. It has several titles, and Title 10 refers to energy. Titles are divided into parts. NRC's regulations are in Title 10 (Parts 0-199). Appendix A to 10 CFR 50 sets out general design criteria (GDC) for nuclear power plants, which set out requirements for the design, manufacture, construction, testing and performance of systems and structures, NRC (1999).

There are 45 GDCs, divided into six categories: 1 - General Requirements; 2 - Protection Against Multiple Barriers for Fission Product Release; 3 - System Protection and Reactivity Control; 4 - Systems Containing Fluids; 5 - Reactor Containment, and 6 - Control of Fuel and Radioactivity.

Appendix B of 10 CFR - Part 50 presents the program requirements for quality assurance. The FSAR contents are established in 10 CFR-Part 50.34 (Contents of Applications; Technical Information). NRC publishes documents called regulatory guides, which, although not mandatory (but strongly recommended), describe methods, standards and acceptable ways to meet the requirements of 10 CFR. These documents are broken down into 10 divisions, where division 1 concerns power reactors.

RG 1.70 (NRC, 1978) establishes the content and format for the FSAR. The Reg Guides mention standards and industry standards that NRC recognizes as safe engineering practices e.g., IEEE Std-323 for electrical and mechanical equipment qualification, IEEE (2004). Some codes and industry standards are considered mandatory and are explicitly mentioned in paragraphs of 10 CFR - Part 50 (eg 10 CFR 50.55a - ASME Code for Pressure Vessels and boilers). See the NRC site (nrc.gov) for details on CFR

Industry standards are prepared by institutions which have began to produce special rules for application in the nuclear area, the main ones being: American Society for Mechanical Engineers (ASME), asme.org; Institute of Electrical and Electronics Engineers (IEEE), ieee.org; American Society for Testing Materials (ASTM), astm.org; Health Physics Society (HPS), hps.org; American Institute of Chemical Engineers (AIChE), aiche.org; Institute of Nuclear Materials Management (INMM), inmm.org.

Technical documents referred as NUREGs are used by NRC in its regulatory action. These reports are diverse in nature and support decision-making. They can result of technical studies, record of experience, training programs, etc. NUREG-0800, Standard Review Plan for Review of Safety Analysis Reports of Nuclear Power Plants is an example. It is used by NRC technical staff for guidance on the assessment of safety analysis reports. Figure 4 displays the general US licensing procedure.

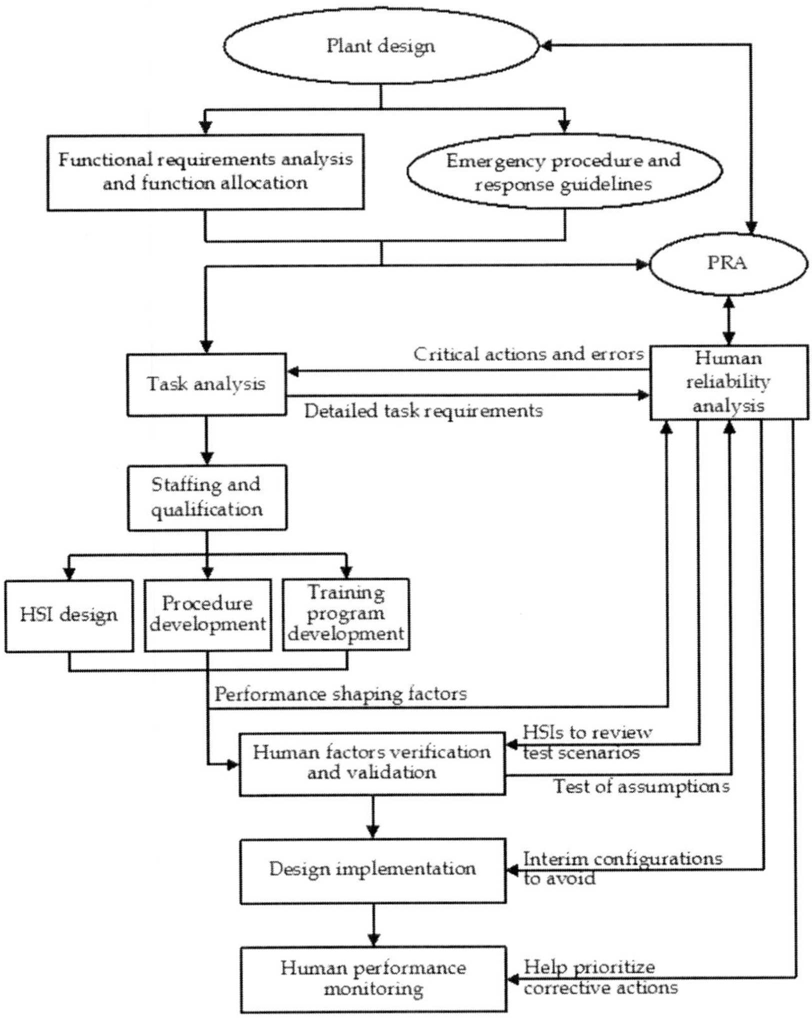

Figure 3. NRC human factor engineering approach (NRC, 2004).

6.9. Licensing in Germany

The Atomic Energy Act (AtG in German) of 1960 provides the legal basis for the peaceful use of nuclear energy in Germany. By the German constitution, states

(*Länder*) are responsible for implementing AtG on behalf of the German federal government. To ensure uniform application of AtG, the Federal Government oversees the states. Section 7 of ATG refers to nuclear installations and their licensing.

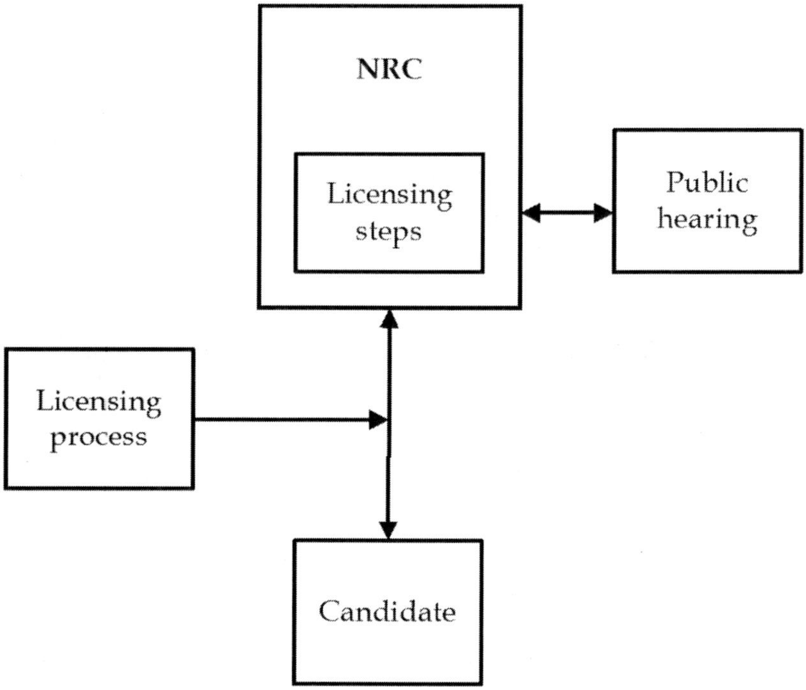

Figure 4. Licensing in US

AtG provisions are supplemented by other laws and regulations of acts in the following areas: radiation protection; environmental impact; emissions control; and service water.

The various acts include the following areas: radiological protection; nuclear licensing procedures; financial insurance; cost of the atomic act; nuclear safety authority; and payment of disposal.

Safety requirements are of general characteristics, providing an environment for different technical solutions, but these solutions must have the same goal of protection. Licensing and supervision authorities have to examine whether this goal is achieved through a variety of safety regulations.

Safety regulations include: a) safety criteria for nuclear power plants, approved by the state committee for nuclear energy; b) BMI (former Ministry of Interior) and BMU (present-day Ministry of Interior and the Environment) guides for qualification of personnel for nuclear power plants; c) safety criteria for final storage; d) safety guidelines of the Committee on Reactor Safety; e) safety standards of the Nuclear Standards Committee; f) standards of the German Institute for Standardization.

The licensee applies for a license to build and operate the plant to the Licensing Authority of the state, preparing the safety report in accordance with the legislation requirements. The state licensing authority examines whether the prerequisites for ensuring the permit were met, assisted by the Organization of Independent Inspection. At the same time, BMU is involved in the process. BMU is assisted by a radiation protection committee. After project evaluation, this committee shall present its recommendations to BMU.

BMU evaluates the recommendations and submit its comments to licensing authorities, which are considered in the decision making process of the state authority. The state authorities, communities near the plant, other authorities and institutions whose areas of responsibility may be affected (nature protection, fire protection, disaster control, etc.) take part in the examination process.

Licensing authorities may request opinions from experts about nuclear safety and radiological protection requirements. However, experts only give technical support to the authorities, having no power of decision in licensing. A step in licensing are public hearings, which may contest the licensing authority, based on current legislation, and consequently taking action to an administrative court.

7. RISK-INFORMED DECISION MAKING

PSA is a methodology that can be applied to provide a structured analysis process to evaluate the frequency and consequences of accidents scenarios in nuclear power plants. NRC first applied PSA in the Reactor Safety Study (NRC, 1975). An important initiative taken by NRC in 1988 was the issuance of Generic Letter GL-88-20, which originated the program known as IPE (Individual Plant Examination). This is because the Reactor Safety Study did not consider each plant individually in the risk assessment.

Since that time, NRC has been using risk assessment and directing the issuance of decisions on complex items associated with or related to safety such as: a) total loss of power (station blackout); b) anticipated transients without reactor shutdown (ATWS); c) pressurized thermal shock events (PTS); and e) Maintenance Rule.

NRC issued the Probabilistic Safety Assessment Policy Statement (NRC, 1995), which incorporated risk assessment as a tool in the regulatory process. It consists of elements that have originated the Risk-informed Decision Making (RIDM) and the Performance Based Regulation (PD).

The following PSA-based RIDM regulatory guides were issued: a) changes in the bases of the specific plant licensing, RG-1.174 (NRC, 2002) ; b) assessment of changes and implementation of technical specifications, RG-1.177 (NRC, 1998c); c) in-service inspections in pipes, RG-1.175 (NRC, 1998a); d) quality assurance, RG 1.176 (NRC, 1998b); e) an approach to determine the technical quality of APS results for RIDM, RG 1.200 (NRC, 2002).

Many of the current regulations, based on deterministic requirements, can not be quickly replaced. In January 2001, Paragraph 69 of the 10 CFR 50 (see nrc.gov), which regulates RIDM, was issued.

'Risk insights' is used to refer to the results and decisions that are made after probabilistic safety assessments are performed. It is necessary to distinguish three approaches or treatments in the decision making process: a) Risk Based (RB); b)) Risk Informed (RI); and c) Performance Based (PB).

The risk-based approach to decision making is the one where only the numerical results of a probabilistic safety assessment are taken into consideration. This causes a strong dependence on the results of risk assessment, due to uncertainties associated with PSA (such as completeness and use of data). NRC does not endorse the risk-based approach, however does not invalidate the use of probabilistic calculations to demonstrate compliance with some criteria.

The risk-informed approach to the process of regulatory decision-making represents a philosophy according to which the outcomes and decisions arising from risk assessment are considered along with other factors to establish requirements that will best target on issues related to the design and operation that impact safety and health of the public.

The RI approach extends and improves the deterministic treatment because it: a) allows explicit consideration of a wide range of changes for safety; b) provides rationale for prioritizing these changes based on risk, operational experience and/or engineering judgment; c) facilitates the consideration of a broad range of resources to support these changes; d) identifies and describes uncertainty sources in the analysis; and e) leads to proper decision making, providing a mechanism to test the results' sensitivity to a set of assumptions.

Where appropriate, a regulatory approach with information on risk can be used to reduce unnecessary conservatism in deterministic treatment, or can be used to identify areas with insufficient conservatism in deterministic analysis and provide the foundation and additional requirements or regulatory actions.

The RI approach lies between the risk-based approach and the purely deterministic treatment. The details of the regulatory approach to be used will determine where the RI-based decision will fall in this spectrum. The concept of defense in depth remains the principle of regulatory practice. The findings and decisions arising from risk assessment can make the elements of defense in depth clearer due to the PSA quantitative approach.

Rules can be either prescriptive or performance based (PB). Prescriptive requirements specify particular aspects, activities or program elements to be included in the project or process, as a means of achieving the desired goal. A performance-based requirement depends on results (measured or calculated, i.e., performance data) to be found. It provides greater flexibility to the licensee to achieve these results.

RIDM philosophy is the reconciliation of the results of PSA insights with the traditional deterministic analysis. Often, PSA results conflict with deterministic insights (defense in depth and safety margin, for example). It is noteworthy that the use of RIDM by the licensee is voluntary.

As a result of policy implementation methodologies for the use of risk information, NRC expected the regulatory process would improve in three aspects: a) by PSA incorporation into regulatory decisions; b) preserving agency's resources; and c) reducing unnecessary effort on licensing.

RIDM follows principles for implementation and evaluation of changes proposed by the licensee, and to evaluate these changes a series of assumptions is adopted by the regulator. It is expected that the proposed changes meet the set of principles described below. PSA techniques can be used to ensure and show compliance with these principles, which are displayed in Table 4.

The evaluation of proposals and licensing acceptance guides adopt these same five principles, according to the eight assumptions detailed next.

Assumption # 1: All safety impact of the proposed change has been assessed in an integrated manner as part of the general approach of risk management, in which the licensee uses risk analysis to improve operational and engineering decisions in the identification of actions to reduce risks, and not to justify the elimination of licensing requirements perceived as undesirable. For those cases where risk increases are proposed, the benefits should be consistent with the increased risk proposal. The approach used to identify changes in requirements must also be used to identify areas where the requirements should be increased or reduced.

Table 4. Principles to be followed by RIDM

Principle	Description
1	Change meets the existing law and is explicitly related to the requested exception or rule change
2	The proposed change is consistent with the philosophy of defense in depth
3	The proposed change has sufficient margins
4	When the proposed change results in an increased frequency of core damage and/or risk, this increase should be small and consistent with the regulations laid down in (51FR30028, 4/8/86)
5	The impact of the proposed change should be monitored using performance measures

Assumption # 2: The content (scope and quality) of engineering analysis (deterministic and probabilistic) performed to conduct and justify the proposed changes have been appropriate to the change nature and scope and should be based on the plant as built and operated, reflecting its operational experience.

Assumption # 3: The plant-specific PSA that supports all licensee proposals has been subject to quality control and an independent evaluation or certification.

Assumption # 4: Consideration of appropriate uncertainties has been provided and decision interpretations supplied, using a monitoring, feedback and corrective actions program to consider significant uncertainties.

Assumption # 5: The use of core damage frequency (CDF) and large early release frequency (LERF) as a basis for PSA acceptance is an acceptable approach to Principle 4.

Assumption # 6: Variations in estimates of CDF and LERF arising from proposed changes to licensing bases will be limited to small increments. Cumulative effects of these changes will be monitored and considered in decision making.

Assumption # 7: Proposal acceptance will be evaluated by licensee in order to ensure that all principles are met.

Assumption # 8: Data, methods and evaluation criteria used to support regulatory decisions should be documented and available for public scrutiny.

Figure 5 displays NRC approach for RIDM.

Regulatory Guide 1.174 (NRC, 2002) describes the approach accepted by NRC to assess the nature and impact of licensing basis conditions (LBC) by considering engineering aspects and application of risk insights.

Regulatory Guide 1.200 describes the approach accepted by NRC to determine that PSA quality, in part or in whole is sufficient to assure its results so that they can be used in regulatory decision making.Figure 6 illustrate the role of the above discussed regulatory guides.

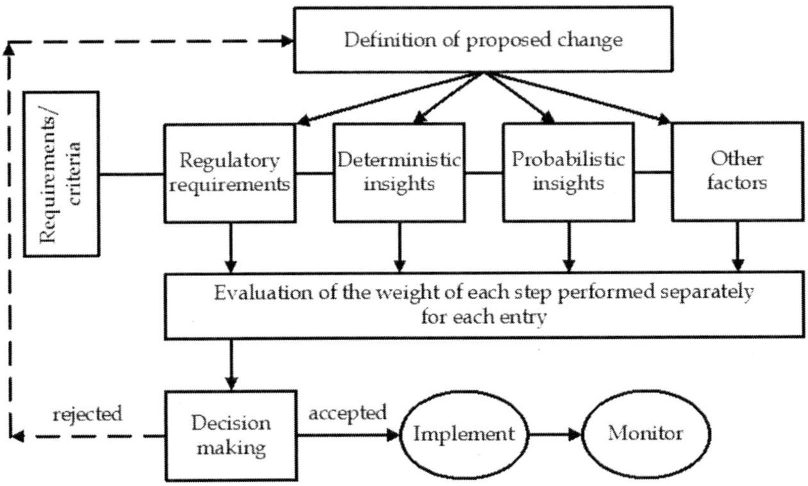

Figure 5. NRC approach for RIDM (NRC, 2002)

In what concerns PSA role in decision making, the key is to provide an assessment of change impacts on risk. It has been necessary to develop a quantitative criterion to serve as a guideline and meet NRC principles. These guidelines have been created to allow comparisons of risk variation evaluations (including internal and external events, full load, low load and shutdown). This criterion uses the core damage frequency obtained from a Level 1 PSA and also the Large Early Release Frequency and is presented in Figure 7.

Kadak & Matsuo (2007) present a discussion about US experience on RIDM usage, where 35 nuclear plants have effectively implemented RIDM. By comparing INPO performance indices with NRC indices for these 35 plants with those of another 19 plants that have not implemented RIDM it is clearly seen that RIDM implementation has significantly improved performance indices. The performance indicators considered in the analysis were: a) Unit capacity factor; b) Automatic unplanned shutdowns (7,000 hr of criticality); c) Safety systems performance; d) Safety injection system actuation; e) Auxiliary feedwater system actuation; f) Power system actuation; g) Fuel element reliability; h) Collective radiation exposure; and i) Water chemistry performance.

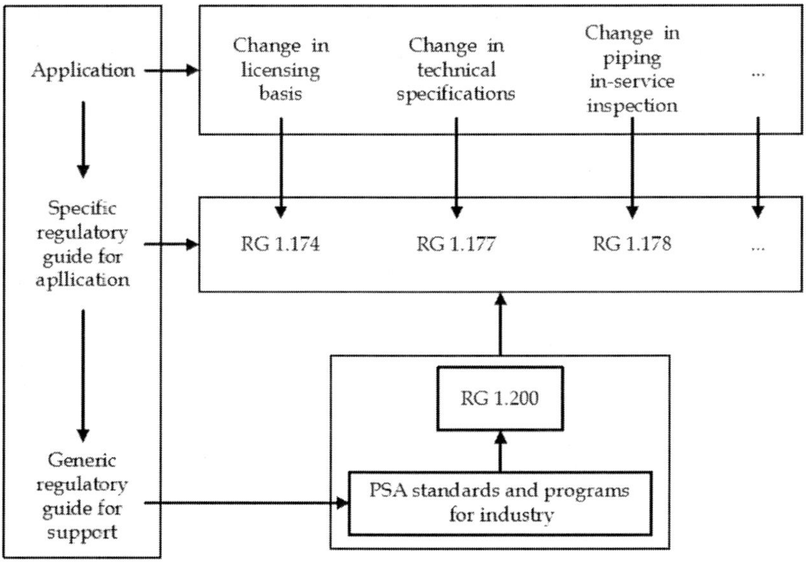

Figure 6. Role of NRC regulatory guides in RIDM.

Table 5 summarizes RIDM status in different European countries (NRC, 2001).

8. FINAL CONSIDERATIONS

It is noted that licensing is characterized by decision making in various fields and disciplines and its steps can influence the course of evolution of the enterprise. The relationship between the licensing agency and applicants for licenses and permits must be honest, keeping the formalism and the independence of the institutions themselves that are recognized with distinct responsibilities. Controversial topics should be discussed openly and the decision making process formalized in writing (technical advice, inspection reports and minutes of meetings).

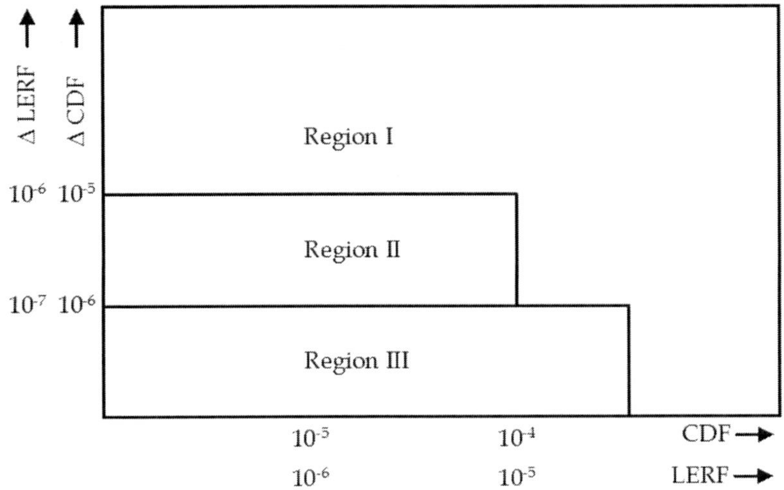

Figure 7. Criteria for the results of a Level 1 PSA (NRC, 2000)

The demonstration that the facility presents no undue risk to the public and the environment may require engineering activities, analytical or experimental, that may significantly affect the initial schedules and preliminary costs estimates. Design features significantly innovative or new reactor concepts require extensive research and testing of prototypes at all levels, such as components, systems, and small-scale facilities.

The physical and mathematical models used in engineering projects are validated by experimental data or calculations. Realistic models of best estimate are preferable to conservative models, but their use requires an adequate treatment of the uncertainties involved, determined from extensively developed databases.

The identification of areas of specialty where the technical staff of the regulatory body lacks experience must occur in a timely manner to enable this qualification or the necessary arrangements to hire consultants or consulting teams.

The timing of any nuclear development should provide the time intervals prior to the granting of licenses and permits for the analysis and evaluation of the regulatory body.

The licensing model in Brazil, in the aspects of safety analysis is deterministic in nature, that is, the plant behavior, after assuming an initiating event or malfunction, is studied with calculation models that describe the physical process of systems reactor.

Table 5. RIDM efforts in somecountries (NRC, 2000)

Country	Status
Spain	Recommends PSA use to define changes in requirements
France	The methodology Optimisation de la Maintenance par la Fiabilité (OMF – Maintenance Optimization for Reliability) has been developed to apply criteria for use of risk information for the optimization of plant maintenance including in-service inspection
Sweden	An update of rules considers in-service inspection use with risk information through qualitative and quantitative approaches
Finland	Expects to rationalize in-service inspection by a combination of deterministic and probabilistic methods under regulator initiative.
UK	Proposals for adoption of methodologies that use risk information in nuclear power plant management represent an evolution.
Switzerland	Expects to produce guidelines for implementation of a risk-informed in-service inspection quantitative method within a few years.
Czech Republic	Believes it is too complicated to introduce a risk-informed approach for in-service inspection, yet its interest on the matter is significant.

The objective of this type of analysis is to check whether the allowed values of key plant variables are exceeded. The probabilistic safety analysis (PSA) focuses on the identification of sequences of events that can lead to meltdown of the reactor, and studies of reliability of safety systems. The objective of this analysis is to indicate potential weaknesses in the design of systems and provide the basis for improving safety.

CNEN has introduced in Standard NE–1.26 (CNEN, 1997) the requirement for risk management, where the operating organization should develop, implement and continuously refine a model for managing the risk associated with various operational configurations. Thus, a probabilistic safety analysis complements the deterministic safety analysis, and it is incorporated into the licensing procedure, because during plant operation, the impact on total risk measured by the model for risk management should be considered. This encompasses decision making involving activities like: a) design modifications and specifications changes or exceptions; b) system configuration management; c) maintenance and testing planning; and d) analysis of operational events.

The responsibility for nuclear safety in all phases of the enterprise belongs to the licensed organization. The licensing activity decides whether the licensee has the technical and organizational competence to fulfill this responsibility.

A very important point in this context concerns the licensing of advanced reactors, like the AP-1000. The risk-informed approach has brought into light the conciliation of deterministic and probabilistic methods for safety analysis. Accident scenarios both in the design basis and also beyond the design basis are being approached much more precisely and many advances in safety philosophy are proving to be effective in this way.

REFERENCES

1. S. K. Ahn, I. S. Kim, K. M. Oh, 2010 Deterministic and risk-informed approaches for safety analysis of advanced reactors: Part I, Deterministic approaches. Reliability Engineering and System Safety, 451458

2. R. C. Borges, F. D'Auria, A. C. Alvim, 2001 Independent qualification of the CIAU tool based on the uncertainty estimate in the prediction of the LOBI test A1-93. Kerntechnik, 66 4 (August 2001), 161170

3. CNEN 1984 Licensing of Nuclear Installations, Standard CNEN-NN-1.04, National Nuclear Energy Commission, Rio de Janeiro, RJ, Brazil

4. CNEN 1997 Safety in Operation of Nuclear Power Plants, Standard CNEN-NN-1.26, National Nuclear Energy Commission, Rio de Janeiro, RJ, Brazil

5. F. D'Auria, G. M. Galassi, 2010 Scaling in Nuclear Reactor System Thermal-hydraulics. Nuclear Engineering and Design, 240 32673293

6. S. Glasstone, A. Sesonske, 1994 Nuclear Reactor Engineering, Reactor Systems Engineering, 2 Chapman & Hall, New York

7. IAEA 1996 Defense in Depth in Nuclear Safety, a Report by the International Nuclear Safety Advisory Group. INSAG-10. International Atomic Energy Agency, Vienna, Austria.

8. IAEA 2000a Safety of Nuclear Power Plants: Design. IAEA Safety Standard Series No. NS-R-1, International Atomic Energy Agency, Vienna, Austria

9. IAEA 2000b Safety of Nuclear Power Plants: Operation. IAEA Safety Standard Series No. NS-R-2, International Atomic Energy Agency, Vienna, Austria

10. IAEA 2009a Deterministic Safety Analysis for Nuclear Power Plants. IAEA Safety Standards, Specific Safety Guide SSG-2 , International Atomic Energy Agency, Vienna, Austria

11. IAEA 2009b Severe Accident Management Programmes for Nuclear Power Plants. IAEA Safety Guide Series No. NS-G-2.15, International Atomic Energy Agency, Vienna, Austria

12. IEEE 2003 Qualifying Class 1E Equipment for Nuclear Power Generating Stations. IEEE Std-323. Institute of Electrical and Electronics Engineers, Piscataway, NJ, USA

13. C. Kadak, T. Matsuo, 2007 The Nuclear Industry's Transition to Risk-informed Regulation and Operation in the United States. Reliability Engineering and System Safety, 92 609618

14. S. Kim, S. K. Ahn, K. M. Oh, 2010 Deterministic and Risk-informed Approaches for Safety Analysis of Advanced Reactors: Part II, Risk-informed Approaches. Reliability Engineering and System Safety, 95 459468

15. R. A. Knief, 1993 Nuclear Engineering, Theory and Technology of Commercial Nuclear Power, Taylor & Francis, Washington, DC, USA

16. Lamarsh, A. Baratta, 2000 Introduction to Nuclear Engineering, Prentice Hall, Upper Saddle River, NJ, USA

17. M. G. Na, S. H. Shin, S. M. Lee, D. W. Jung, S. P. Kim, J. H. Jeong, B. C. Lee, 2004 Prediction of Major Transient Scenarios for Severe Accidents of Nuclear Power Plants. IEEE Transactions on Nuclear Science, 51 2 April 2004), 313321

18. NRC 1975 Reactor Safety Study- An Assessment of Accident Risks in US Commercial Nuclear Power Plants. WASH-1400, NUREG-75/014, Nuclear Regulatory Commission, Washington, DC, USA

19. NRC 1978 Standard Format and Content of Safety Analysis Reports for Nuclear Power Plants. RG-1.70, Nuclear Regulatory Commission, Washington, DC, USA

20. NRC 1995 Use of Probabilistic Risk Assessment Methods in Nuclear Regulatory Activities, Final Policy Statement. Federal Regulation-60FR 42622, Nuclear Regulatory Commission, Washington, DC, USA

21. NRC 1996 Standard Review Plan for the Review of Safety Analysis Reports for Nuclear Power Plants, LWR Edition. NUREG-0800, Nuclear Regulatory Commission, Washington, DC, USA.

22. NRC 1998a An Approach for Plant-Specific, Risk-Informed Decision Making: Inservice Testing Assurance. RG-1.175, Nuclear Regulatory Commission, Washington, DC, USA

23. NRC 1998b An Approach for Plant-Specific, Risk-Informed Decision Making: Graded Quality Assurance. RG-1.176, Nuclear Regulatory Commission, Washington, DC, USA

24. NRC 1998c An Approach for Plant-Specific, Risk-Informed Decision Making: Technical Specifications. RG-1.177, Nuclear Regulatory Commission, Washington, DC, USA

25. NRC 1999 General Design Criteria for Nuclear Power Plants. Appendix A of 10 CFR 50. Nuclear Regulatory Commission, Washington, DC, USA

26. NRC 2001 Technical Committee Meeting on Risk Informed Decision Making (RIDM). Nuclear Regulatory Commission, Washington, DC, USA

27. NRC 2002 An Approach for Using PRA in Risk-Informed Decisions on Plant-Specific Changes to the Licensing Basis. RG-1.174, Nuclear Regulatory Commission, Washington, DC, USA

28. NRC 2004 Human Factors Engineering Review Program. NUREG-0711, Nuclear Regulatory Commission, Washington, DC, USA

29. NRC 2007 An Approach for Determining the Technical Adequacy of Probabilistic Risk Assessment Results for Risk-Informed Activities. RG-1.200, Nuclear Regulatory Commission, Washington, DC, USA

30. G. Petrangeli, 2006 Nuclear Safety. Butterworth-Heinemann/Elsevier, Amsterdam, The Netherlands

Chapter 4

Radiological Releases and Environmental Monitoring at Commercial Nuclear Power Plants

Jason T. Harris[1]

[1] Idaho State University, United States of America

1. INTRODUCTION

The generation of electricity from nuclear power has become increasingly important due to the growing concerns of global climate change. Nuclear energy has long been recognized as a leading energy source that produces minimal pollution to the environment that can contribute to this phenomenon. In addition, nuclear power offers an attractive option for countries looking for energy source diversification. Currently there are 442 commercial nuclear power reactors operating in the world (International Atomic Energy Agency [IAEA], 2010, 2011). These power plants contribute about 19% of the electricity production today. The United States of America (U.S.) has the largest commercial nuclear reactor fleet in the world with 104 operating reactors (U.S. Nuclear Regulatory Commission [USNRC], 1996). Of these reactors, 69 are pressurized water reactors (PWRs) and 35 are boiling water reactors (BWRs), located on 65 sites around the country. These power plants contribute about 20% of the U.S. electricity production.

Although it is known that commercial nuclear power plants release small amounts of radioactivity into the environment, there is still the potential for these releases to impact public health. This is especially important today as changes are occurring in nuclear power plant operations including: higher electric generating capacities, increased power levels due to mechanical uprates, and plant life extensions. Public health effects must be reexamined as new light water reactor designs are being considered for construction. In addition, recent events at multiple nuclear power plants in the U.S. involving unplanned releases, especially tritium (^{3}H), have led to increased scrutiny on monitoring and evaluating releases. Changes in radiation protection recommendations and regulations also warrant further and continued investigations in these matters. Although Harris (2007) and Harris & Miller (2008) have performed numerous studies of nuclear power effluent releases and environmental monitoring, data collection and analysis must continue to be performed for the entire nuclear industry.

This chapter focuses on recent research that has been conducted in the areas of commercial nuclear power radiological releases and environmental monitoring by the author. Although the emphasis will be on studies performed in the United States of America, international comparisons will be made where appropriate.

2. BACKGROUND

Commercial nuclear power plants release small amounts of radiation into the environment under normal operating conditions. Many of the radioactive isotopes that are released are in the form of gaseous or liquid effluents and solid radioactive waste conditioned by the plant. These releases represent some of the by-products of electrical energy generation (Eisenbud & Gesell, 1997).

Three categories of radioactive by-products are produced during routine operation of a commercial light-water reactor: fission products, neutron activation products, and tritium (Glasstone & Jordan, 1980). Fission products are created as a result of the radioactive decay of the nuclear fuel. Approximately 300 different nuclides are formed in the operating reactor. Most of these nuclides are radioactive. Although there is a large quantity of fission products formed, many have little impact on the radioactive releases to the environment because of their extremely short half-lives (<1 day), small quantities, or biological insignificance. Gaseous fission products important to these releases include:^{3}H, ^{85}Kr, and ^{133}Xe. Iodine, solid at room temperature, is also released as a gaseous effluent due to vaporization. Important dose significant iodine isotopes include: ^{131}I, ^{133}I, and ^{135}I. Other decay daughters of produced fission products may also appear in the gaseous effluents as particulate matter (USNRC, 1976a, 1976b).

Activation products are formed by neutron interactions with oxygen in water and air, with nitrogen and argon in air, and with impurity corrosion elements. Like fission products, many of the neutron activation products produced are insignificant in reactor effluents due to their short half-lives (<1 day) or small quantities. Relevant gaseous activation products include: ^{13}N, ^{14}C, ^{16}N, and ^{41}Ar (NCRP, 1985, 1987). Important liquid and solid waste activation products arising from interaction of neutrons with corrosion and erosion elements include: ^{51}Cr, ^{58}Co, ^{60}Co, and ^{59}Fe (Kahn, 1980; USNRC, 1976a,1976b).

Tritium (^{3}H or T), is produced as a result of both nuclear fission (ternary fission) and neutron activation of deuterium (^{2}H). Tritium is typically treated separately because it is produced in such large quantities compared to any other effluent nuclide and because it arises from other nuclear reactions. One significant source of tritium is the interaction of high energy neutrons with boron. Boron is used in PWRs for shim control (as boric acid) and BWRs as a burnable poison (Glasstone & Jordan, 1980). Tritium is also formed from the interaction of neutrons with ^{6}Li (as lithium hydroxide in water treatment).

Typically, the radioactive emissions from operating nuclear power reactors result in insignificant doses to the general population. In 1988, when 110 nuclear power plants were operating at 70 sites in the United States, the mean collective effective dose commitment from all pathways ranged from a low of 1.1×10^{-5} person-Sv (0.0011 person-rem) to a high of 0.16 person-Sv (16 person-rem). The collective dose commitment for the 150 million persons living within the 2-80-km annuli was 0.75 person-Sv (75 person-rem) for that year (USNRC 1995). Other studies performed throughout the world have shown similar results for population doses around nuclear power plants (Walmsley et a.l, 1991; Ziqiang et al., 1996; Kim & Han, 1999; Nedveckaite et a.l, 2000; Liu et al., 2003; Quindos Poncela et al., 2003).Harris (2007) performed a study to look at the doses for maximally exposed individuals from all plants. A review of epidemiological studies of cancer in populations near nuclear facilities showed that in all scientific reports analyzing nuclear power plants, a cause and effect relationship between cancer risk and radiation exposure could not be found (Patrick, 1977; Jablon et al., 1991; Shleien et al., 1991;Lopez-Abente et al., 1999).

There has been a gradual reduction in both liquid and gaseous emissions from power reactors due to improvements in fuel performance and radioactive waste treatment system technology (Harris, 2002). However, the Electric Power Research Institute (EPRI) reports that although radioactive isotopes captured by these systems reduce effluent quantities, radioactive solid waste volumes increase (2003). Also, with longer operating times and license extensions, the accumulation of spent fuel is becoming more important. Many plants have begun storing spent fuel on-site in independent storage facilities. The ageing of existing nuclear power facilities and the increasing accumulation of radioactive wastes have led to an increased emphasis on solid radioactive waste disposal However, at this time doses to the public have not increased during the handling or transportation of radioactive waste shipments. Worldwide estimates also show that nuclear power will continue to grow and thus remain a source of radioactivity exposure to the public.

2.1. Regulatory Criteria of Releases

The principles that apply to U.S. nuclear power plant radiological releases include consensus scientific recommendations, governmental regulations (Code of Federal Regulations [CFR]), and specific criteria in each plant's operating license. Dose limits, concepts and models based on scientific agreement about radiation effects are recommended by the International Commission on Radiological Protection (ICRP) and the NCRP. Government radiation protection guidance is developed by the U.S. Environmental Protection Agency (USEPA) and approved by the President to assist federal agencies, such as the U.S. Nuclear Regulatory Commission (USNRC), in developing radiation protection regulations. This guidance is usually in agreement with the ICRP or the NCRP. The regulatory standards developed are then required to be incorporated into each nuclear power plant as radiological effluent technical specifications (RETS) that are to be followed through procedures and programs (Andersen, 1995).

Since the inception of nuclear power, federal radiation protection regulations have been based upon the recommendations of the ICRP. The initial ICRP recommendations, published as ICRP Publication 1 and ICRP Publication 2, provided dose limits, models, and radiation concepts. Subsequent to these initial recommendations, the ICRP issued three major revisions, ICRP Publication 26, ICRP Publication 60, and ICRP Publication 103. These recommendations lowered the annual dose limits for members of the public and revised dose models and concepts. ICRP Publication 26 (1977) recommended an annual dose limit of 5 mSv y^{-1} (0.5 rem y^{-1}) to critical members of the general population (pregnant women and children). Critical members of the general population are those that are more susceptible to radiation effects. ICRP Publication 60 (1991) lowered recommended annual dose limits further to 1 mSv y^{-1} (0.1 rem y^{-1}) for members of the general population. ICRP Publication 103 (2007) continues with this dose limit. Another important recommendation in terms of reactor releases is given in ICRP Publication 29. This document provides the Committee's recommendations for evaluating pathways between radioactive materials released into the environment and man (ICRP, 1978).

One important recommendation made by the NCRP, published as NCRP Report 92, is specifically concerned with public radiation exposure resulting from nuclear power (NCRP, 1987). The report outlines dose concepts, risks, and technical information regarding the nuclear fuel cycle. In 2011, NCRP will release another recommendation (Report 169) on effluent and environmental monitoring design. Other organizations, including the United Nations Scientific Committee on the Effects of Atomic Radiation (UNSCEAR), have had a tremendous influence on the understanding of radiation concepts. The UNSCEAR reports yearly to the General Assembly and periodically issues (every four to five years) the important publication, Sources and Effects of Ionizing Radiation (2000). Another United Nations organization, the International Atomic Energy Agency (IAEA), also influences the practice of radiation protection and issues radiation related reports.

The updated regulations important to nuclear power plant radiological effluents are found in 40 CFR 190 (USEPA 1977). These regulations include limits on radiation doses received by members of the public off-site of the nuclear power plant. During normal operation, the annual dose to any member of the public shall be limited to: 0.025 mSv y^{-1} (25 mrem y^{-1}) to the whole (total) body; 0.075 mSv y^{-1} (75 mrem y^{-1}) to the thyroid; and 0.025 mSv y^{-1} (25 mrem y^{-1}) to any other organ. The USEPA has also set forth guidelines for the maximum amount of radioactivity released into the environment (e.g. 5 mCi of ^{129}I per gigawatt-year of electrical energy produced).

The USNRC issues standards and regulations for radiation protection and nuclear plant operations. Standards for radiation protection are contained in 10 CFR 20 (USNRC, 1991). These standards incorporate the dose concepts and models from the older ICRP Publication 26 and 40 CFR 190. The criteria in 10 CFR 20 regarding dose limitations include: a public dose limit of 1 mSv y^{-1} (0.1 rem y^{-1}), compliance with USEPA's 40 CFR 190 standards, and a requirement for a licensee survey of radiation levels in unrestricted areas, in controlled areas, and in effluent releases. Appendix B to 10 CFR 20 includes limits on effluent concentrations for radiological releases in air and water. These concentration limits are derived from occupational inhalation and ingestion annual limits on intake (ALIs) adjusted to reflect the dose limits set forth by the standards.

USNRC standards for nuclear power plant operations are contained in 10 CFR 50. These standards include criteria for radiological effluent technical specifications, effluent release design objectives and limits, and notification and reporting for events involving the release of radioactive materials. Technical specifications on effluents from nuclear power plants are listed in 10 CFR 50.36a (USNRC, 1996). The specifications require that the licensee comply with 10 CFR 20; that procedures be established and followed regarding the control of effluents; that a radioactive waste treatment system be installed, maintained, and used; that a report be submitted annually to the USNRC regarding effluent releases and the attributed estimated doses to the public; and that procedures be developed that comply with the principle of achieving radiation levels ALARA. Appendix I to 10 CFR 50 gives numerical guides for design objectives and limiting conditions for operation to meet the ALARA criterion for radiological effluents. Doses to members of the general public from radioactive material in liquid effluents released to unrestricted areas shall be limited to 0.003 mSv y^{-1} (3 mrem y^{-1}) to the whole (total) body, and 0.010 mSv y^{-1} (10 mrem y^{-1}) to any other organ. The air dose due to the release of noble gases in gaseous effluents is restricted to 0.010 mGy y^{-1} (10 mrad y^{-1}) for gamma radiation, and 0.020 mGy y^{-1} (20 mrad y^{-1}) for beta radiation. The public dose from ^{131}I, ^{3}H, and all particulate radionuclides with half-lives greater than eight days in gaseous effluents is limited to 0.015 mSv y^{-1}(15 mrem y^{-1}) to any organ. Standards in 10 CFR 50 also cover notification in the event of an abnormal radiological release.

Criteria for nuclear power plant effluents are contained in the radiological effluent technical specifications (RETS), which are part of the nuclear power plant operating license. The RETS include the Limiting Condition for Operation (LCO). The LCO is a description of the criteria that are to be met, the conditions under which the criteria apply, the actions to be taken if criteria are not met, and surveillance requirements to demonstrate that the criteria have been met. The RETS must also contain a site specific Offsite Dose Calculation Manual (ODCM). The ODCM contains both the methodology and parameters used in calculating offsite doses resulting from radiological effluents and the REMP. The USNRC Regulatory Guide 4.1 outlines the programs for monitoring radioactivity in the environs of nuclear power plants (USNRC, 1975). The RETS and ODCM must be approved by the USNRC as part of the license application and approval process. Radiological effluent technical specifications guidelines are contained in NUREG-0133 (USNRC, 1978). The annual effluent report covers plant operations from the previous calendar year. The report includes a summary of the quantities of radiological effluents and solids discharged by the plant. USNRC Regulatory Guide 1.112 aids nuclear power plants in calculating effluent releases.

2.2. Environmental Monitoring

Prior to the issuance of a construction permit or an operating license for a nuclear power station, federal agencies (i.e. USNRC) are required to assess the potential environmental effects of that facility to ensure that issuance of the permit or license will be consistent with the national environmental goals prescribed by the National Environmental Policy Act (NEPA) of 1969 and the Federal Water Pollution Control Act. In order to obtain information needed for this assessment, applicants are required to submit a report on the potential environmental impacts of the station and associated facilities. After the station becomes operational, an annual environmental report must be submitted to ensure continued compliance of the requirements set forth in the facility's license and of the Acts stated previously.

Radiological environmental monitoring programs at nuclear power plants are required in accordance with the Code of Federal Regulations. Development and maintenance of these programs are under the guidance of several federal documents. These radiological environmental monitoring programs are established to monitor the radiological impact of reactor operations on the environment. Objectives of these programs include: identification, measurement and evaluation of existing radionuclides in the environs of the facility and fluctuations in radioactivity levels which may occur; evaluation of the measurements to determine the impact of operations on the local radiation environment; collection of data to refine radiation transport models; verification that radioactive material containment systems are functioning to minimize environmental releases to levels that are as low as reasonably achievable (ALARA) and; demonstration of compliance with regulations. Implicit in these objectives are the requirements to trend and assess radiation exposure rates and radioactivity concentrations in the environment that may contribute to radiation exposures to the public. The results of the REMP are submitted as part of the plant's annual environmental report.

Each plant establishes their own, unique REMP program to reflect site-specific conditions and surrounding population characteristics. The program consists of preoperational and operational components. The preoperational program is conducted in part to measure background levels and their variations in environmental media in the area surrounding the plant. Environmental media include: milk produced from cows or goats, broadleaf vegetation, fish, fruits and vegetables, edible aquatic invertebrates, surface water, drinking water and ground water.

Each plant is to also make changes to its REMP program as conditions change. But, it has been reported recently that many plants are decreasing their programs due to budget constraints and lack of positive radioactivity measurements. This reduction can lead to decreased litigation protection, decreased public confidence, and potential unreported or undetected releases. Reduced REMP programs have led to recent public opinion and regulatory problems for several facilities due to unexpected and/or undetected tritium releases. Decreased lower limits of detection (LLDs) and minimal detectable activities (MDAs) reportedly have led to newly quantifiable low levels of many radionuclides in the environs around nuclear power stations. Changes in operating conditions may also lead to new radionuclide transport pathways being developed, as has been seen with precipitation scavenging and concentration in ice.

Ultimately, a nuclear power plant's REMP program is designed to assess the impact of radiological releases on the environment and the public. Public opinion of the nuclear power industry has traditionally been very troubled, especially with the accidents at Three-Mile Island and Chernobyl. Positive public opinion to nuclear power can only be achieved through truthfulness by the nuclear power company regarding operations and radiological releases and accurate and comprehensive monitoring of these releases.

3. EFFLUENT RELEASE STUDY

As commercial nuclear power electrical generation steadily increases in the U.S. and the rest of the world, it has become even more important to evaluate the release of radioactive materials into the environment. An easy way to track industry wide effluent releases is by performing trend analyses. Accumulated data may also be used for analysing reactor power up-rate consequences, protecting the nuclear power industry against litigation, and for assisting in new power plant siting. Most importantly, collecting and maintaining an effluent database is necessary in maintaining a favourable public perception regarding the low environmental and biological impact of nuclear power. This is especially important now as several recent, inadvertent releases of radioactive materials from nuclear power plants have occurred. Because of these circumstances, the author has compiled and analysed the effluent data for all U.S. commercial nuclear power plants since 1995. Presented here is also an update of the comprehensive study performed by Harris & Miller (2008).

The classification and monitoring of liquid and gaseous radiological releases is fairly uniform around the world. The classification is based on the nuclide, chemical or physical form, and dose or activity significance. In the U.S., gaseous effluents are divided into fission and activation gases, iodines, particulates (with half-lives greater than eight days), and tritium. Liquid effluents are divided into fission and activation products, dissolved and entrained gases, tritium, and gross alpha activity. International organizations and other nations use the same categories, but combine the fission and activation products and the dissolved and entrained gases in liquid effluents (UNSCEAR, 2000; Harris, 2002).

The classification of radioactive releases is important because dose calculations are based upon them. For example, the collective effective doses calculated by UNSCEAR (2000) use these effluent categories. The groupings also allow plants and nations to compare and benchmark with another. Unlike the simplified general UNSCEAR model, the USNRC model requires specific nuclide, meteorological, and site specific conditions. Hence, this model provides more accurate estimates of dose.

Other studies have been performed to assess the doses from nuclear power radiological releases. Vold (1983) determined the ratio of the collective effective dose equivalent (CEDE) via a specified ingestion pathway relative to that CEDE by inhalation per annual releases of a radionuclide. Kim & Han (1999) and Liu et al. (2003) assessed the impact of tritium released from nuclear power plants in China. Both of these studies confirmed that the doses were less than 1% of the regulatory limits.Ziqiang et al. (1996) reported similar results not only for tritium, but for other radionuclides as well.

What is very common in the nuclear power industry is trending and benchmarking of data. This is done to improve plant operations and management. Many organizations that oversee different aspects of nuclear power plants use these methods for comparison. These comparisons may be advantageous or detrimental to a plant. For example, with radiological releases, high activities compared to other plants can lead to lower profits due to higher premiums from American Nuclear Insurers (ANI). It can also lead to scrutiny from the Institute of Nuclear Power Operations (INPO) and greater surveillance from USNRC. Thus, these comparisons are very important. Gilbert (1994) identifies statistical analyses suitable for detecting trends in environmental contamination data. Accurate trend analyses can aid plants in these aforementioned areas. Trend analyses were performed for the data over the 15 year period using the Mann-Kendall non-parametric test. Inspections of release trends over the fifteen year period help identify areas of concern with these releases. Future estimates of release radioactivity and public doses can then be made from these analyses.

3.1. Methodology

The data utilized for the effluent release study were taken from the annual radioactive release reports provided by the nuclear power plants to the USNRC as required in their operating license conditions. These reports were either provided directly to the author from the licensee or taken from the USNRC Agency wide Documents Access and Management System (ADAMS). The reports provide categorical effluent release data, nuclide specific radioactivity, and site specific data needed for dose calculations. Population information not provided by the licensees was taken from appropriate census reports (U.S. Census Bureau, 2010).

Data was analysed for those reactors that have operated for the 15 year period of 1994-2009. This length of time is long enough to allow plants to stabilize in the event of long shutdown periods and allows evaluation of plants for at least seven refuelling cycles. Events that may affect releases, such as power-uprates and failed fuel from defects, will also show up in this period. The beginning of this data set also coincides with the cessation of tracking radiological effluents by the U.S. in 1994. In this time frame, 103 reactors were operating. Browns Ferry Unit 3 began operation in 2006, to become the 104th operating reactor.

Effluent radioactivity was obtained from data reported by the nuclear power plants in their annual radioactive material release reports. The effluent data was compiled for all operating PWR and BWR plants from 1995 - 2005. The completeness of the data was 98%. In keeping with U.S. nuclear power effluent report formatting, data was compiled and analysed using the same categories as those listed in USNRC Regulatory Guide 1.21. The four gaseous effluent categories used were: fission and activation gases (F/A), total iodine (^{131}I), particulate matter or particulates, and tritium. The three liquid effluent categories used were: fission and activation products, dissolved and entrained gases, and tritium. Because the radioactivity levels of the fission and activation products and dissolved and entrained gases are several orders of magnitude smaller than tritium, those two categories were added together and listed as "F/D". This category replicates the reporting done by UNSCEAR. Gross alpha radioactivity was not included in this study.

Trend analyses were performed for the data over the time period using the Mann-Kendall non-parametric test. This procedure was used since missing values were allowed and the data need not conform to any particular distribution (Gilbert, 1994). Inspection of trends over the time period identifies the overall direction of industry effluent releases and can roughly be used to predict future releases.

For this updated study, one dose assessment methodology was used. The collective effective doses (CED) were calculated for the U.S. population using the UNSCEAR methodology. For these dose calculations, the effluent data was normalized. This was achieved by taking the amount of radionuclides released per unit of electrical energy generated each year. This method is the most common way to normalize effluent data. The electrical energy generated per year was obtained by multiplying the net electrical energy generated by the capacity factor. Capacity factor is defined as the gross electricity generated divided by the product of the licensed capacity and reference time. Normalizing data in this manner takes into account the operational performance of the nuclear power plant. However, it also assumes that effluent release amounts are a direct consequence of operation time. The author cautions against making simplistic comparisons of radioactive releases with the electrical energy generated because of the many factors which affect the amount of radioactive materials released, including the condition of the fuel, primary system integrity, design of effluent and radioactive waste treatment systems, maintenance activities, operations, and equipment performance.

3.2. Results and Discussion

3.2.1. Radiological Effluent Releases

The annual variation of total nuclear power plant radioactivity released in gaseous effluents in PWRs and BWRs are shown in Figs. 1 and 2, respectively. As expected, the activity from PWR releases is higher than that from BWRs due to the greater number of plants. Regardless of this fact, the average tritium release from PWRs is also higher due to chemistry practices that create more tritium in the plant. Nearly every category from both reactor types is fairly level in terms of activity released for the entire time period. The evaluation of the data over the time period partly eliminates variations in annual values. The advantage of using 15 years of data is that operation anomalies, such as long shutdown times for maintenance, are averaged out. The one notable exception appears in the PWR particulate category. In 2003, one plant experienced an annual release over five orders of magnitude above the mean. This single event was significant enough to skew the entire industry release activity, especially since the annual radioactivity released in particulate matter is so low compared to tritium or fission and activation gases. The increase in 2005 was due to higher activity releases by several plants.

The annual variation of total nuclear power plant radioactivity released in liquid effluents in PWRs and BWRs are shown together in Fig. 3. As expected, and for reasons similar to that of the gaseous releases, PWR liquid radioactivity in releases is higher than in BWRs. Liquid releases have stayed very constant over the 11-year period. The most notable exception is the pronounced decline in BWR non-tritium (F/D) radioactivity from 2003 - 2007.

3.2.2. Radiological Effluent Trends

U.S. industry effluent trends were evaluated using the Mann-Kendall non-parametric test. The Mann-Kendall test was performed as follows: For any given release category, x, its feature vector consists of the release summation from all plants appearing in a given year i. These release activities are ordered from the first year, 1995, to the final year, 2009, that data was gathered.

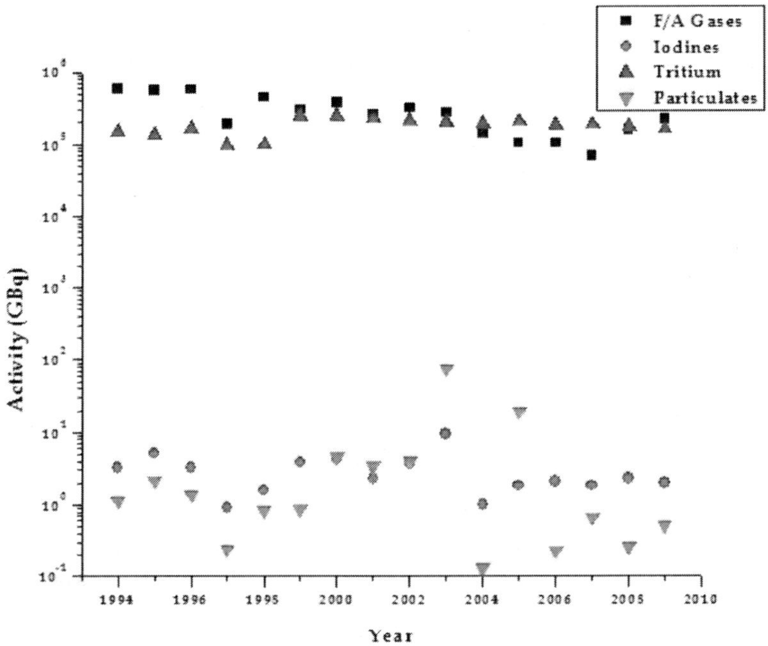

Figure 1. Variation of total radionuclide activity released in gaseous effluents from PWR plants

The x value for each year is compared to all other years greater than that year, and Kendall's statistic S is calculated as follows:

$$S = \sum_{k=1}^{n-1} \sum_{j=k+1}^{n} \operatorname{sgn}(x_j - x_k) \tag{1}$$

where:

$$(x_j - x_k) = \begin{cases} 1, & if\, x_j - x_k > 0 \\ 0, & if\, x_j - x_k = 0 \\ -1, & if\, x_j - x_k < 0 \end{cases}$$

Generally, if a dataset displays a consistently increasing or decreasing trend, S will be positive or negative, respectively, with a larger magnitude indicating the trend is more consistent in its direction. By using the *sgn* function, the algorithm used was able to detect trends featuring either large or small increase steps from year to year equally. The S statistic is then compared to the corresponding P-value (Hollander & Wolfe, 1999).

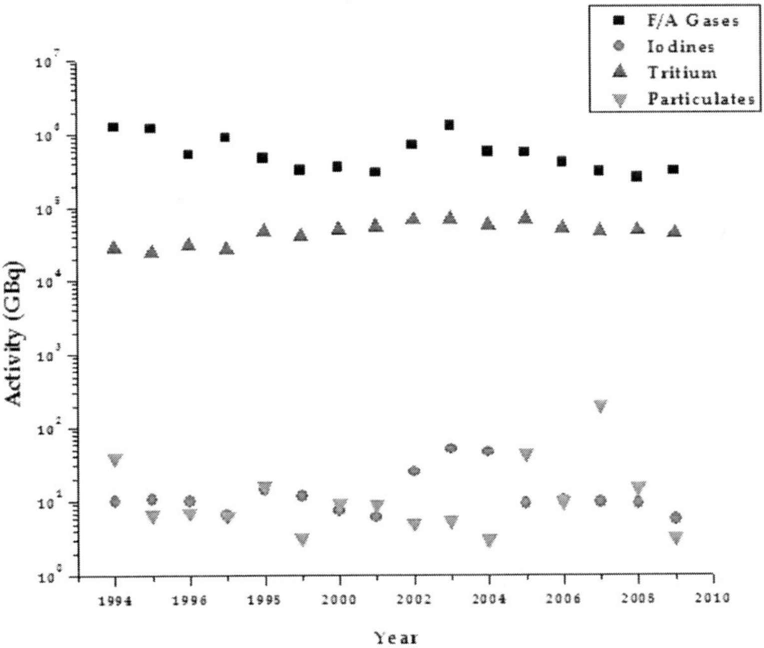

Figure 2. Variation of total radionuclide activity released in gaseous effluents from BWR plants

Under the null hypothesis, H_0, that there is no trend displayed by the time series, the distribution of S is then expected to have a zero-mean and variance. The Mann-Kendall test was performed on BWR plant releases, PWR plant releases, and all plant releases (BWR and PWR combined) at a significance level of 0.05.

The results of the trend test are shown in Table 1. Over the past 15 years, it can be seen that for most effluent categories, the releases are level, meaning there is no increasing or decreasing trend. For these categories, improvements in radioactive waste treatment and reactor operations are offset by increased power production, increased capacity factors, and power up-rates. Looking at PWRs only, gaseous fission and activation products and liquid F/D have decreased while liquid tritium has increased. Reduction in the fission and activation products over the years is probably a direct result of longer holdup times for radioactive decay. In addition lower fission and activation product activities may be due to improved fuel performance from better manufacturing methods. Also, longer operation times in recent years provide stability to the reactor, leading to less fuel shock and defects. The increasing trend for liquid tritium activity is due to an increase in liquid discharges by these plants. Formerly zero-discharge plants have begun to release liquids again to avoid build-up of their tritium inventory. Coupled with this practice, over the last fifteen years many plants have reduced recycling of boron for reactor control. This procedure contributes to increased tritium production.

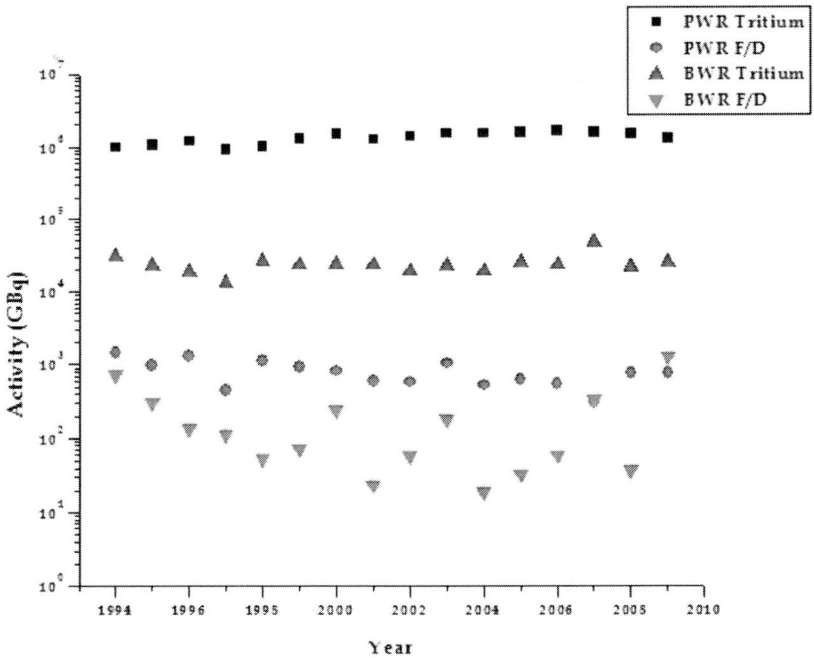

Figure 3. Variation of total radionuclide activity released in liquid effluents from PWR and BWR plants

For BWRs, liquid tritium and gaseous releases have increased and gaseous fission and activation releases have decreased. The increase in gaseous and liquid tritium are relatively new phenomenon and is probably related more to the increased power production and capacity factors over the last several years than anything else. Looking at the entire industry (PWRs and BWRs combined), all effluent releases except gaseous fission and activation products and iodine and liquid tritium are level. Because PWRs make up 66% of the U.S. industry, their releases have a greater impact on the overall release trends, as is evident with the increasing liquid tritium.

Table 1. Mann-Kendall trend results for U.S. commercial nuclear power plant radiological effluent releases from 1995 – 2009

Release Category	S Statistic	Trend
PWR Gaseous--F/A gases	-61	Decreasing
PWR Gaseous--Iodines	-15	None
PWR Gaseous--Tritium	-3	None
PWR Gaseous--Particulate	-11	None
PWR Liquids--Tritium	57	Increasing
PWR Liquids – F/D	-35	Decreasing
BWR Gaseous--F/A gases	-41	Decreasing
BWR Gaseous--Iodines	-13	None
BWR Gaseous--Tritium	31	Increasing
BWR Gaseous--Particulate	7	None
BWR Liquids--Tritium	27	Increasing
BWR Liquids – F/D	-7	None
Total Gaseous--F/A gases	-65	Decreasing
Total Gaseous--Iodines	-21	Decreasing
Total Gaseous--Tritium	-3	None
Total Gaseous--Particulate	13	None
Total Liquids--Tritium	57	Increasing
Total Liquids – F/D	-13	None

3.2.3. Radiological Impact of Effluent Releases

Tracking effluent release quantities is important in determining radioactivity levels in the environment. However, dose determination of the effluents must be performed to estimate the human effects of these radiation sources. The collective effective dose (CED) from radiological effluent releases was obtained using an average collective dose calculation method used by UNSCEAR.UNSCEAR (2000) calculates population dose by first calculating the CED per unit release of radionuclides released from reactors and then normalizes the value with the electrical energy generated. The CED is divided according to type of release (airborne or liquid), radionuclide category (noble gases, tritium, C-14, iodine, and particulate matter), and pathway (immersion, ingestion, inhalation, and external irradiation. The normalized collective effective dose model is given by:

$$D_{CE} = \sum_i \frac{A_i}{E} D_i \quad (2)$$

where: A_i =activity of release category i (GBq); D_{CE} = total CED (person-Sv GW^{-1} y^{-1}); D_i = collective dose for release category i (person-Sv PBq^{-1}); and E =energy produced by the nuclear reactor (GW y^{-1}).

The dose assessment procedures for this model are applied to a model site with representative environmental conditions. The average population density used is 20 km^{-2} within 2,000 km of the site. Within 50 km of the site, the population density is taken to be 400 km^{-2}. These parameters were obtained from previous UNSCEAR assessments and take into account the transport and dilution of released radionuclides from nuclear installations. The parameters used are assumed to not underestimate dose. Using this model site, the collective effective dose per unit release is obtained for the different release categories. Due to its lack of specificity, this model should be used for general comparisons only. The collective effective doses estimated from commercial nuclear power plant radiological effluent releases are very low especially when compared to other man-made sources of radiation. The doses only represent up to a few percent of the regulatory limits (Harris & Miller, 2008).

From the collective effective doses, effective doses were computed for the entire U.S. population to give average annual doses. This was done by taking the CEDs calculated for each release category, gaseous and liquid, and dividing them by the U.S. population for each year (U.S. Census Bureau, 2010). The effective doses were than summed to give a total dose for each person. The results of these effective doses are given in Table 2. For the 15-year period, total effective doses ranged from 5.42×10^{-8} mSv (5.42×10^{-6} mrem) to 1.68×10^{-7} mSv (1.68×10^{-5} mrem). The doses to an average person would be expected to be even lower since many do not live near a nuclear power plant. This is just one of several ways to calculate a very general effective dose for the population.

4. U.S. REMP EVALUATION

An evaluation of all U.S. nuclear power plant radiological environmental monitoring programs (REMP) was conducted from 1995-2007. An attempt was made to assess the significance of the radionuclides detected in the environment compared to natural and other man-made radiation sources. It is important to note that detected concentrations of radionuclides in the local environment as a result of nuclear power radiological releases are very low. The percentage of plants sampling different pathways, the percent cumulative exposures, and differences between control and indicator measurements were determined.

4.1. Methods

Inspection and analysis of the industry REMP data was taken from the annual summary tables from each nuclear power plant's annual radiological environmental monitoring report. The summary data includes the following parameters: sample medium, type and number of analyses performed, required (LLD), the mean and range of the positive measured values of the indicator locations, the mean, range, and location of the highest indicator annual mean, the mean and range of the positive measured values of the control locations, and the number of non-routine reports sent to the USNRC. In this evaluation, the highest indicator values were compared to the control values.

The following environmental pathways and sample analyses (in parentheses) were investigated: bottom/shoreline sediment (gamma spectral analysis), fish (gamma spectral analysis), edible aquatic invertebrates (gamma spectral analysis), surface water (gamma spectral analysis, gross beta, and ^{3}H), drinking or potable water (gamma spectral analysis, gross beta, and ^{3}H), ground water (gamma spectral analysis, gross beta, and ^{3}H), air particulate filters (gamma spectral analysis and gross beta), airborne radioiodine (^{131}I), milk (^{90}Sr and ^{131}I), food products such as fruits and vegetables (gamma spectral analysis), broadleaf vegetation (gamma spectral analysis), soil (gamma spectral analysis), grass (gamma spectral analysis), and direct radiation using thermoluminescent dosimeters (TLDs). In addition, pathways not normally sampled by most plants, like precipitation, were also studied.

For all data, only positive measured values were used in the statistical calculations. During the study period there were over 1.6×10^{6} analyses performed on environmental media collected as part of the required monitoring programs. Broken down, this averages to about 1.27×10^{5} analyses y^{-1} and 2.0×10^{3} analyses $site^{-1}\ y^{-1}$.

4.2. Results and Discussion

4.2.1. Cumulative Dose Contribution from Effluent Release Pathways

The percent contribution by each pathway to the public was determined using previously complied effluent data. The calculations provided by each plant from their REMP data identified the most important pathways for their respective sites. Fig. 4 shows the mean results for the seven most frequently sampled pathways or media for all plants. The greatest dose contributor comes from direct radiation released by the plant. This is especially true for BWR plants, which have no secondary loop for the reactor produced steam.

The water, milk, and sediment media, at about the same level, provide the next highest percent contributions to dose. Water media, which includes surface, ground, and drinking types, may contribute to dose to a much larger extent depending on the plant location. Those that are on saltwater sites may not have a known drinking water pathway, or dilution may be so great that submersion doses are low or non-existent. On the other hand, plants on freshwater sites typically have drinking water pathways and less dilution of their effluents (with the exception of sites located on the Great Lakes). All other pathways contribute to a much lesser extent.

4.2.2. Plant Pathway Sampling

The investigation of the number of pathways sampled by each plant is important to gauge the diversity and scope of each plant's REMP in relation to each other. Used in conjunction with the results from Section 4.3.1, omissions of pathways can be reconsidered based on the overall percent contribution of dose. Obviously, programs will vary greatly since the power plant sites themselves also vary greatly. For example, plants may be located on coastal sites, in river valleys, in arid locations, or on small, man-made bodies of water. The level of human activity around the plant will also influence the degree of surveillance. The number of media sampled by plants ranged from five to more than 20, with a mean of 11.

The results of the percentage of plants sampling different media in their REMP are displayed in Fig. 5. As expected, 100% of plants sample for air particulate matter, ^{131}I and direct radiation. Direct radiation is important as it is the greatest dose contributor to members of the public, especially from BWR plants. Milk, once sampled by nearly all plants, is now sampled by only 52% of plants. This is due to the reduction of the milk pathway (from cows or goats) in areas around plants. In lieu of milk sampling, plants may substitute broadleaf vegetation. One important characteristic of the data is that nearly 56% of plants sample groundwater. Recent unplanned tritium releases went unnoticed at several plants due to the lack of groundwater monitoring at those sites. In part due to this study and a national industry initiative, all plants began a groundwater monitoring program starting in 2006. At the time of this research and writing, updated data on nuclear power plant groundwater monitoring was not available. The dose important fish pathway is sampled by 88% of plants. Edible aquatic invertebrates are sampled by only 19% of the plants, mainly due to the lack of the pathway.

Table 2. Average effective doses received by members of the public in the U.S. from commercial nuclear power plant radiological effluent releases

Year	Electrical Energy Produced (GW)	U.S. Population (x 10^3)	Annual Effective Dose (mSv GW^{-1} $person^{-1}$)						
			Gaseous Releases				Liquid Releases		Total
			F/A Gases	Total I-131	Tritium	Particulates	Tritium	F/D Gases	
1995	77.1	266,557	8.36×10^{-8}	1.95×10^{-10}	1.68×10^{-8}	1.28×10^{-9}	2.93×10^{-8}	2.90×10^{-8}	1.60×10^{-7}
1996	77.3	269,667	7.79×10^{-8}	2.75×10^{-10}	1.31×10^{-8}	1.10×10^{-9}	3.18×10^{-8}	2.89×10^{-8}	1.53×10^{-7}
1997	71.9	272,912	1.08×10^{-7}	1.29×10^{-10}	1.90×10^{-8}	1.47×10^{-9}	2.71×10^{-6}	1.22×10^{-8}	1.68×10^{-7}
1998	74.9	276,115	1.38×10^{-8}	2.80×10^{-10}	1.46×10^{-8}	2.66×10^{-9}	2.68×10^{-6}	1.37×10^{-8}	7.19×10^{-8}
1999	82.3	279,295	7.00×10^{-9}	1.75×10^{-10}	1.57×10^{-8}	3.06×10^{-10}	2.83×10^{-6}	1.10×10^{-8}	6.24×10^{-8}
2000	85.2	282,402	7.98×10^{-9}	1.80×10^{-10}	1.48×10^{-8}	1.08×10^{-9}	3.05×10^{-6}	1.07×10^{-8}	6.53×10^{-8}
2001	87.8	285,329	5.58×10^{-9}	9.21×10^{-11}	1.50×10^{-8}	8.57×10^{-10}	2.54×10^{-6}	7.97×10^{-9}	5.49×10^{-8}
2002	88.6	288,173	8.42×10^{-9}	1.95×10^{-10}	1.73×10^{-6}	6.62×10^{-10}	2.70×10^{-8}	1.96×10^{-8}	7.32×10^{-8}
2003	87.0	291,028	1.44×10^{-6}	3.79×10^{-10}	1.51×10^{-6}	3.04×10^{-9}	2.87×10^{-8}	1.15×10^{-8}	7.30×10^{-8}
2004	88.1	293,907	6.94×10^{-9}	2.67×10^{-10}	1.39×10^{-6}	2.07×10^{-10}	2.64×10^{-8}	6.38×10^{-9}	5.42×10^{-8}
2005	88.6	295,753	7.49×10^{-9}	9.78×10^{-11}	1.59×10^{-6}	5.11×10^{-9}	2.77×10^{-8}	5.99×10^{-9}	6.23×10^{-8}
2006	89.3	298,593	5.40×10^{-9}	9.84×10^{-11}	1.28×10^{-8}	4.24×10^{-10}	2.96×10^{-8}	6.18×10^{-9}	5.45×10^{-8}
2007	88.9	301,580	4.82×10^{-9}	1.09×10^{-10}	1.13×10^{-8}	9.48×10^{-9}	2.61×10^{-8}	4.95×10^{-9}	5.67×10^{-8}
2008	88.9	304,375	4.44×10^{-9}	1.03×10^{-10}	1.21×10^{-8}	1.85×10^{-9}	2.82×10^{-8}	1.01×10^{-8}	5.68×10^{-8}
2009	86.8	307,007	5.77×10^{-9}	7.31×10^{-11}	1.14×10^{-8}	3.32×10^{-10}	2.41×10^{-8}	1.41×10^{-8}	5.58×10^{-8}

4.2.3. Remp Summary

Numerical results for the REMP summary are not listed here due to the breath of the study. Comparisons are only made between the control samples and the indicator or test measurements. Direct radiation pathways include radiation from buildings and plant structures, airborne material that might be released from the plant, cosmic radiation, "fallout", and the naturally occurring radioactive materials in soil, air and water. Analysis of the TLDs, which are used by all plants, indicated that there were no increased radiation levels attributable to plant operations. The mean BWR exposure was slightly higher than that for PWR plants, but the control locations actually gave higher mean values for all plant types.

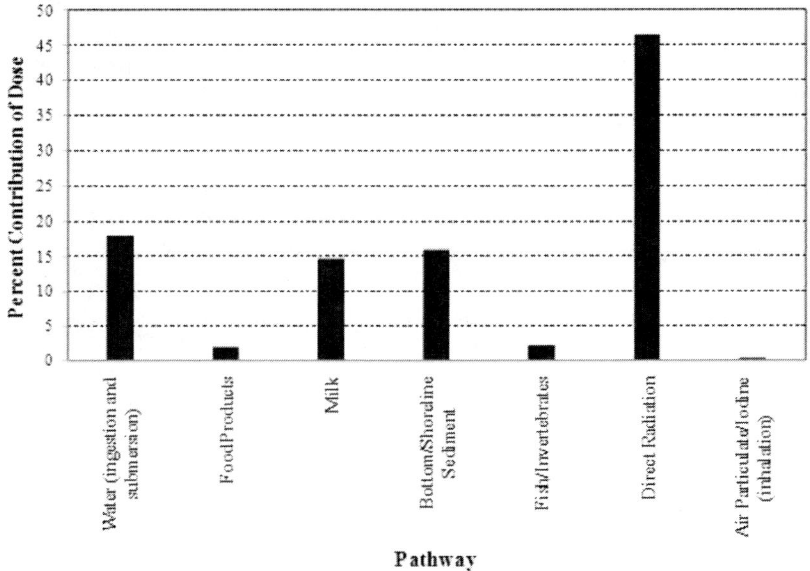

Figure 4. Percent total cumulative dose contribution of various pathways resulting from U.S. nuclear power plant effluent releases

Airborne pathway measurements include analyses of ^{131}I, gross beta, and gamma spectra. For all ^{131}I measurements, the values were below the LLD. Therefore, the reported mean and median measurements are for half of the LLD values. For the entire 13 years, only one plant reported an indicator value above the LLD. The gross beta and gamma spectral analyses showed no difference between the control and indicator stations. These airborne pathway measurements indicated that there was no increased radioactivity attributable to plant operations.

Terrestrial pathways that include milk, vegetation, grass, and food products, all gave calculated mean indicator measurements that were below that of the controls, Similarly for all waterborne pathways (drinking water, ground water, surface water, fish and invertebrates, and sediments), measurements of the media indicated that there was no adverse radiological impact to the surrounding environment attributed to plant operations for the study period. Individually, several plants have had measurements that resulted in indicator values being significantly higher than those from control stations.

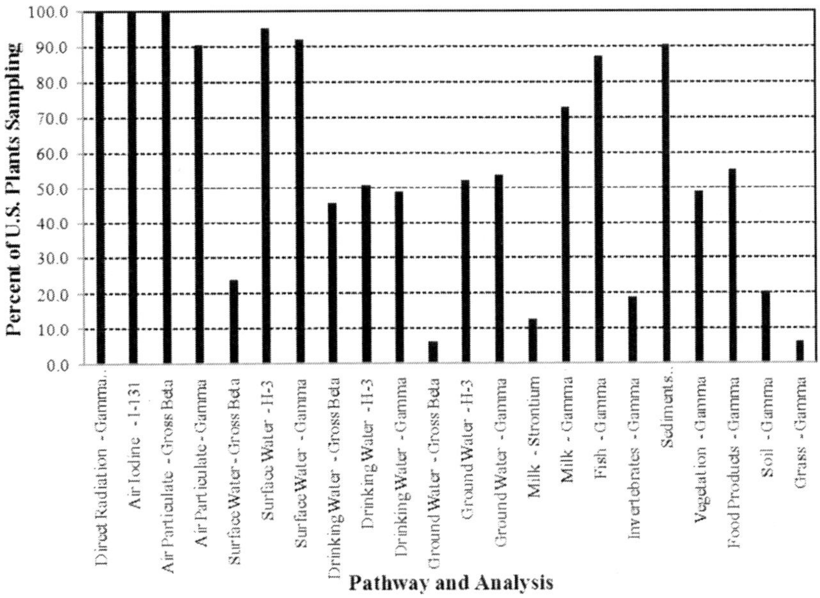

Figure 5. **Percent of U.S. nuclear power plants sampling different REMP pathways and performing specific analyses**

The summary data strongly suggest that for the great majority (>99.9%) of the analyses, the control and indicator concentrations were indistinguishable. Potential plant produced radionuclides were detected in both indicator and control locations. In the past 13 years, the concentration of radionuclides in the environment has stayed nearly level. Prior to this time, radionuclide concentrations showed significant downward trends, mainly due to the decay of "fallout" radionuclides, evidenced in the historical data provided by the plants. The overall trend of the REMP data for these plants is "de minimis" levels of anthropogenic radioactivity with occasional samples showing radioactivity above the LLD. These data also indicate no correlation between total dose and distance from the plants.

The analytical results from the REMP study demonstrate that the routine operation of all facilities had no significant or measurable radiological impact to the environment for the study time period. Also, these environmental surveillance programs continued to demonstrate that the doses to members of the public as a result of nuclear plant operations remained significantly below the federally required dose limits specified in 10 CFR 20 and 40 CFR 190.

5. TRITIUM RECAPTURE STUDY

Tritium is an isotope of hydrogen that decays with a half-life of 12.3 y to helium (^{3}He), while emitting a low energy beta particle (18.6 keV). In nuclear reactors, tritium is formed in two different ways: by ternary fission of uranium, and by activation (Luykx & Fraser, 1986). Produced tritium that ultimately ends up in the nuclear reactor coolant can then be released into the environment through waste discharge in either gaseous or liquid forms (Kim & Han, 1999). Because tritium is generated in very large volumes, it is the predominant gaseous and liquid effluent released by these reactors.

Although tritium is not considered to be a particularly toxic radionuclide (Hill and Johnson 1993) and releases by commercial nuclear power plants have traditionally been well below regulatory limits, control and monitoring is important because of sensitive public concerns regarding radioactivity releases (Andersen 1995; Liu et al 2003). This is especially vital at the present time because recent, unplanned tritium releases have occurred at several nuclear power plants in the U.S. These unplanned releases may result in additional dose both to occupational workers and members of the public. This has led to increased scrutiny by plant owners and government agencies. In 2007, Harris also reported that the largest number of non-routine environmental sampling results was from tritium in surface water.

This section describes research conducted at the Cook Nuclear Plant (U.S.) to investigate the behaviour of tritium released in airborne effluents. In addition to studying the behaviour and movement of tritium in the environment, the study has been conducted to help develop a sampling and analysis protocol for other plants to adopt. Methods and results are primarily taken from Hinchcliffe, 2010.

5.1. Background

At the Cook Nuclear Plant, a site consisting of two PWRs, an event in 2007 set off an investigation into tritium release on the site. The plant is located in the state of Michigan along the eastern shore of Lake Michigan. In May of that year sampling began on the site as a result of a leak containing tritium found outside the auxiliary building. This sampling resulted in elevated tritium levels being discovered in the north storm drain outfall, which drains into Lake Michigan. This outfall had previously been sampled on a semi-annual basis with no previous indication of elevated tritium levels (Harris et al., 2008). The investigation concluded that tritium originating in the spent fuel pool was leaving the monitored vents on top of the containment buildings of both units and contributing to the elevated levels. Findings led to increased sampling of precipitation, air-conditioning condensate, surface and well water, and frost formed in refrigerators/freezers (Harris et al. 2008). The importance of precipitation was noted in the form of washout in which tritium can be scavenged by falling raindrops and tritium vapour can exchange with ordinary water in the atmosphere (Chamberlain & Eggleton, 1964; Tokuyama & Oonishi, 1997; Harris et al., 2008). In this way, precipitation becomes contaminated with tritium and sampling can be used to calculate a washout coefficient. This coefficient allows for estimation of tritium washout due to the precipitation which can be used to estimate the extent of tritium recapture on site property (Harris et al., 2008).

After an initial study in 2007 it was decided that additional sampling should be performed to determine the variations in recapture on the site as the weather changed through the year, and that snow should be sampled during the winter. Snow was sampled to determine if there was a difference in washout between rain and snow and to determine if recapture of tritium occurred differently with snow than it did with rain.

5.2. Methodology

The investigation of tritium washout was performed by taking samples at pre-determined site locations and analysing the collected samples for tritium. For rain collection, gauges with a surface area of 7.85×10^{-4} m^2 were placed 0.25 m above the ground surface. Thirty two (32) rain gauges were placed in each land-based 22.5° sector of the plant site. Additional gauges were located in the predominant seasonal wind directions; SW in the summer and N and NE in the winter. The rain samples from this investigation were collected between 22 October 2007 and 21 October 2008 during 13 rain events. Snow was collected on 28 March 2008, 11 February 2009 through 16 February 2009, and 5 January 2010 through 10 March 2010.

Samples were taken both inside and outside the plant Protected Area, as shown in Fig. 6 and 7. The sample locations varied in distance from the containment stacks, ranging from 5 m (location 12) to 1.5 km (location 32). For control purposes, location 16 was placed 24 km from the release point. Snow was collected immediately after snow fall events, sampled from the surface of existing snow piles, and sampled with cores to evaluate depth differences from existing piles. Surface snow samples were obtained using 20 mL collection cups. Core samples yielded four to six, 10 cm sections, depending on pile depth. Ground level concentrations of tritium were also determined for calculation of a new washout coefficient.

Sample analysis was done using a liquid scintillation counter (Model 2910TR, PerkinElmer). The samples were prepared by mixing 5 mL to 10 mL of the sample with 10 mL of scintillation cocktail in 20 mL vials and counting for 30 minutes. The tritium counting efficiency was 60%, and the lower limit of detection (LLD) was 30 Bq (819 pCi).

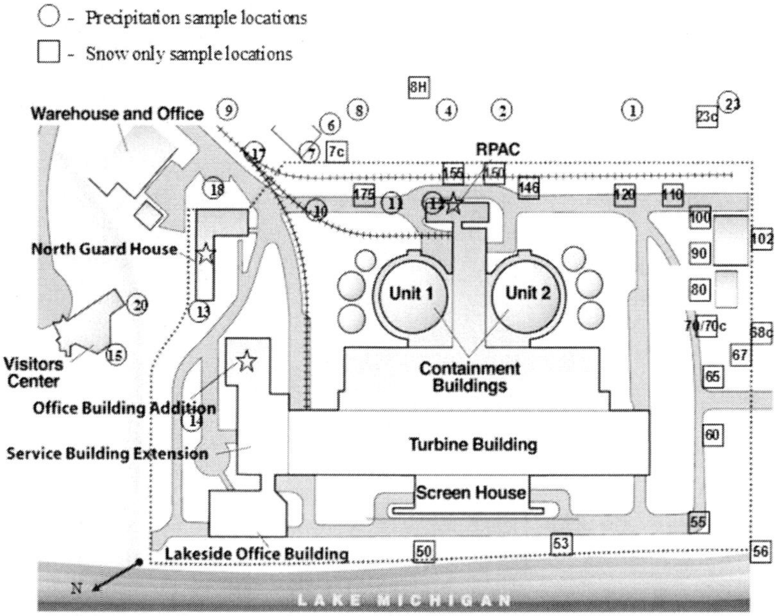

Figure 6. Rain and snow sample locations located outside the Cook Nuclear Plant protected area

5.3. Results

5.3.1. Tritium Concentrations

In total, 23 of the 32 rain locations had a positive result for tritium at some point in the study. With the snow, 44 of 56 sample locations resulted in a tritium concentration above the LLD. The highest rain concentration was 514

Bq L-1 (13,900 pCi L^{-1}), and occurred at location 18, 130 m NE of the Unit 1 containment vent. The highest snow concentration was 1,014 Bq L^{-1} (27,400 pCi L^{-1}), and occurred at snow sample location 110, 140 m SSE of the Unit 1 containment vent. The highest concentrations of tritium in the snow occurred in the old snow that had accumulated in piles on site. The positive results were found to match well with the meteorological data, with positive samples collected in the predominant wind direction.

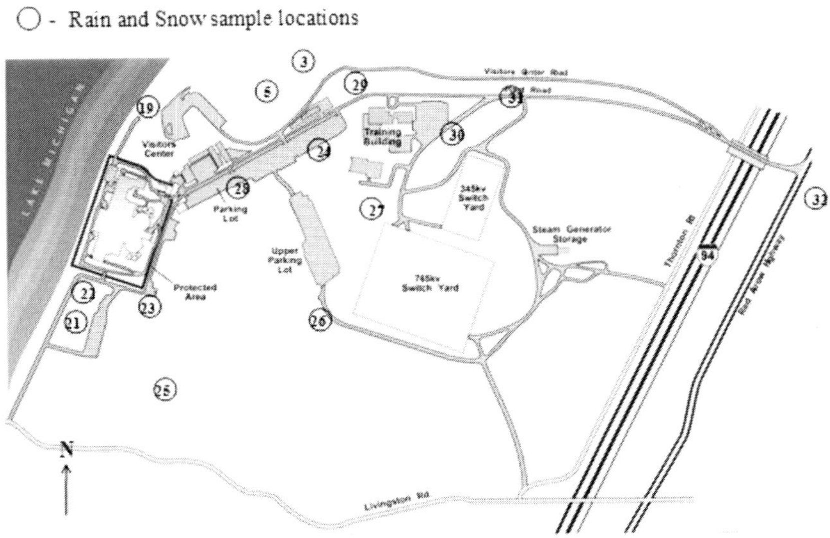

Figure 7. Sample locations for rain and snow within and immediately outside the Cook Nuclear Plant protected area

There were a total of five core samples taken and all indicated the presence of tritium in at least one core piece. Four of the five core samples showed an increasing concentration of tritium closer to the surface of the snow pile. The only core sample that did not show this pattern was core sample V which was taken from the visitor's parking lot, 200 m NNE from the Unit 1 containment vent. All core pieces in this sample were below LLD, with the exception of core piece "2" (second piece from the bottom of the core).

5.3.2. Washout

A rain washout coefficient was calculated from the measured tritium concentrations found in the samples, the atmospheric tritium level, and the meteorological data gathered at the Cook Nuclear Plant. The washout coefficient is calculated by:

$$\Lambda = \frac{\omega}{\chi_0 \times H_{eff}} \tag{4}$$

where: Λ = washout coefficient (s^{-1}); ω = tritium deposition rate (Bq m^{-2} s^{-1}); χ_0 = atmospheric tritium concentration at ground-level (Bq m^{-3}); and H_{eff} = effective height (m). The effective height is calculated using the dispersion equation (Chamberlain & Eggleton, 1964):

$$H_{eff} = \frac{\sum_s \frac{Q}{\sqrt{2\pi}\sigma_{ys} U_S}}{X_{0,cal}} \tag{5}$$

where: Q = tritium emission rate (Bq m^{-2} s^{-1}); σ_{ys} = standard deviation of distribution of concentration in the y direction (m); U_s= mean wind speed (m s^{-1}); $X_{0,cal}$ = mean concentration for the ground-level atmospheric concentration (m). The subscript S refers to the atmospheric stability. Using the local meteorological conditions, the deposition rate is given as:

$$\omega = \frac{C \times I_R}{3600} \tag{6}$$

where: C = tritium concentration in rainwater (Bq L^{-1}); and I_R = mean rainfall intensity(mm h^{-1}).

The emission rates were calculated using the continuous tritium release rates from the Cook Nuclear Plant Annual Radioactive Effluent Release Reports (ARERR) for the periods on which the precipitation events occurred. The precipitation intensities were calculated for each precipitation event, with an average value of 3.1 mm h^{-1} for rain, and 1.1 mm h^{-1} for the snow. For the rain samples the washout coefficient values ranged from 7.84 x 10-7 s^{-1} to 1.13 x 10-4 s^{-1} with an average washout coefficient of (2.04 ± 1.85) x 10-5 s^{-1}. Washout coefficients were only calculated for the fresh snow. The reason for this as that as the snow sits, it accumulates tritium from the atmosphere, and thus skews the value of the washout coefficient. The fresh snow was analysed shortly after the snow event. The calculated washout coefficients for the fresh snow ranged from 2.21 x 10^{-6} s^{-1} to 2.33 x 10^{-5} s^{-1} with an average value of (1.30 ± 0.75) x 10-

5 s^{-1}. Upon comparing the rain and snow washout coefficient sample populations using a two sample t-test, it was determined that there was a significant difference between the washout coefficients in the snow and rain at the 0.05 decision rule level.

5.4. Discussion

Several other studies have reported values of washout coefficients in the literature. A theoretical washout coefficient based on the rate of exchange of tritiated water (HTO) vapour by rain, was calculated to be 10^{-4} s^{-1} (Chamberlain & Eggleton, 1964). In Japan, Tokuyama & Oonishi (1997) used a mean rainfall intensity of 2 mm h^{-1} and calculated a washout coefficient from releases at a nuclear plant to be $(7.3 \pm 4.1) \times 10^{-5}$ s^{-1}. The previous study at Cook Nuclear Plant calculated a washout coefficient of $(9.70 \pm 8.40) \times 10^{-5}$ s^{-1}, based on a rainfall intensity of 6.2 mm h^{-1} (Harris et al., 2008). The rain washout coefficient in this study is smaller than those previously listed, but on the same order of magnitude as Tokuyama & Oonishi and Harris et al. The release rates used for this study were monthly averages rather than instantaneous rates used in several of the other studies. In addition, the calculated effective heights were calculated higher in this study than the other studies that listed the effective height used.

Annual fluctuations of weather have multiple effects on the process of tritium recapture at nuclear sites. The locations of greatest recapture vary between seasons and even within particular seasons. Snow recapture at the Cook Nuclear Plant occurred in locations of the site that previously were not known for rain recapture, thus presenting new pathways of tritium movement needing to be studied. Snow also presents a unique problem, as it is not removed from the site at the same rate that rain would be removed, thus allowing for tritium to be absorbed into snow piles that accumulate around the site. This allows for snow piles to obtain tritium concentrations far above what simple washout can cause in precipitation alone. The piles also appear to accumulate a higher concentration of tritium on and near the surface of the pile where the snow is exposed to the atmosphere.

The results here suggest that individual sites will have to come up with their own sampling methods based on individual site characteristics such as annual fluctuations in wind and precipitation. Variables will change from site to site, and long term annual studies will need to be performed to fully comprehend the behaviour of tritium recapture. A single industry model would be unable to account for the many individual variables that would change between sites. Studies are on-going at Cook Nuclear Plant to better understand tritium behaviour.

6. FUTURE RESEARCH

Although the research presented here provided important insights into commercial nuclear power plant discharges, more studies are needed to truly understand effluent trends and nuclear power plant radioactivity. As long as commercial nuclear power plants continue to operate and release radioactivity into the environment, there will always be a need to monitor, track, and evaluate these releases. This is especially true as new power plants are constructed and existing plants make operational changes. Evolutions in standards, recommendations, and regulations may also warrant studies to address the potential impact. One expansive area of research that is developing comes from recommendations made by the ICRP in Publications 103 and 108 (2007, 2009). Specifically there is mention of radiological protection of the environment and the development of reference plants and animals. If nations adopt these recommendations, large scale studies at nuclear power plants may need to be undertaken.

In the U.S., a number of potential new areas of research involving nuclear power radiological releases are being investigated. First, revisions in 2009 of USNRC Regulatory Guides 1.21 and 4.1 have prompted plants to begin monitoring their ^{14}C emissions. Only a few plants had done this prior to the release of these revisions and now several organizations are evaluating methods and procedures to best accomplish this. Other countries are also evaluating this scenario, if they are not already monitoring for ^{14}C. In conjunction with the tritium washout study presented in Section 5, several plants are conducting their own studies on tritium releases and movement. The majority of plants in the U.S. are now monitoring groundwater and studying its movement and transport of radionuclides. Finally, with the construction and operation of new plants, studies on effluent fate and transport, dose assessment and environmental impact are sure to be conducted.

7. CONCLUSION

In conclusion, the research presented here will hopefully provide stakeholders with some additional insights on assessing nuclear power plant radiological releases and environmental monitoring programs. Evaluation of radionuclide impacts in terms of potential increased dose to people, in relation to natural background, is necessary to determine the true significance of any detection. Maintained effluent tracking and analysis will continue to be important for the future of the commercial nuclear power industry. Effluent trending may also reveal insight into the effect of increased reactor lifetime operation on radioactivity releases. Currently, many plants have been approved for, or are applying for, operating license extensions. Tracking data for siting of new nuclear power plants may also be used to determine environmental radionuclide build-up and long-term nuclear power health effects.

Although the research presented here showed that normal nuclear power operations have no significant effect on the environment, continued surveillance is important. The potential development of new pathways may require changes in monitoring techniques. Doses to members of the general public, and even occupational workers, may increase because of these pathways. Finally, continued industry analysis of radiological effluent releases and environmental monitoring are important for maintaining favourable public opinion about nuclear power. The more accurate, scientifically based information that citizens can be provided with, the more likely they are to make informed, non-emotional judgments about nuclear power.

ACKNOWLEDGEMENTS

This chapter was written from research results and material developed by the author over the last decade. I wish to thank the many friends, family, and colleagues, too numerous to mention by name, for their contributions. Special appreciation goes out to Dr. David Miller, Dr. Richard Brey, Dr. Thomas Gesell, Dr. George Sandison, Dr. George Miley, and the late Dr. Herman Cember. Their guidance and support have truly helped shape not only my career, but my life as well. I would also like to thank Dr. Kenneth Sejkora for sharing his knowledge in nuclear reactor aspects of environmental monitoring and effluents. I am also indebted to my students, especially William Hinchcliffe, Gavin Hawkley, and Peter Lee, for their assistance in many parts of the research presented in this chapter. Lastly, and most importantly, I would like to thank my wife, Dr. Maria Okuniewski. If it were not for her constant support, encouragement, and love, I would not be where I am today.

REFERENCES

1. R. Andersen, 1995Criteria for radiological releases, In: Radiation Protection at Nuclear Reactors, C. Maletskos, (Ed.), 21 44MPP, 978-0-94483-855-6Madison, USA

2. A. Chamberlain, A. Eggleton, 1964Washout of tritiated water vapour by rain. J Air Wat Poll, 8 135 149

3. Eisenbud, M. & Gesell, T. (1997). Environmental Radioactivity (4th), Academic Press, ISBN 978-012-2351-54-9, San Diego, USA

4. Electric Power Research Institute 2003 Strategies for managing liquid effluents-options, actions, and results, EPRI, Palo Alto, USA

5. S. Glasstone, W. Jordan, 1980 Nuclear Power and its Environmental Effects, American Nuclear Society, 978-0-89448-022-5La Grange Park, USA

6. R. Gilbert, 1994 Statistical Methods for Environmental Pollution Monitoring, Van Nostrand Reinhold, 978-0-47128-878-7New York, USA

7. J. Harris, 2002 Comparative Study of Commercial Nuclear Power Plant Radiological Effluents, University of Illinois, Urbana, USA

8. J. Harris, 2007 Public Health Analysis Resulting from Commercial Nuclear Power Plant Radiological Emissions, Purdue University, West Lafayette, USA

9. Harris, J. & Miller, D. (2008). Radiological effluents released by U.S. commercial nuclear power plants from 1995-2005. Health Phys, 96 734 743

10. J. Harris, D. Miller, D. Foster, 2008Tritium behavior at a nuclear power reactor due to airborne releases. Health Phys, 95 203 212

11. R. Hill, J. Johnson, 1993Metabolism and dosimetry of tritium. Health Phys 65 628 647

12. W. Hinchcliffe, 2010 Investigation of Tritium Recapture at Cook Nuclear Power Plant from Airborne Effluent Releases, Idaho State University, Pocatello, USA

13. M. Hollander, D. Wolfe, 1999 Nonparametric Statistical Methods, Wiley, 978-0-47119-045-5New York, USA

14. IAEA (2010). Nuclear Power Reactors in the World (2010), IAEA, Vienna, Austria978-9-20105-610-8

15. IAEA (2011). Nuclear Power Plant Information, In: IAEA PRIS Database, 10.03.2011, Available from: http://www.iaea.org/cgi-bin/db.page.pl/pris.charts.htm

16. ICRP 1977 Recommendations of the ICRP- ICRP Publication 26, Pergamon Press, 978-0-85951-080-6Oxford, England

17. ICRP 1978 Radionuclide Release into the Environment: Assessment of Doses to Man- ICRP Publication 29, Pergamon Press, 978-0-08032-335-0Oxford, England

18. ICRP (1991). 1990 Recommendations of the ICRP-ICRP Publication 60, Pergamon Press, 978-008-0411-44-6, Oxford, England

19. ICRP (2007). 2007 Recommendations of the ICRP- ICRP Publication 103, Pergamon Press, 978-070-2030-48-2, Oxford, England

20. ICRP 2009 Environmental Protection: the Concept and use of Reference Animals and Plants- ICRP Publication 108, Pergamon Press, 978-0-44452-934-3Oxford, England

21. S. Jablon, et al. 1991Cancer in populations living near nuclear facilities, a survey of mortality nationwide and incidence in two states. J Am Med Assoc 265 1403 1408

22. B. Kahn, 1980Composition and measurement of radionuclides in liquid effluent from nuclear power stations, In: Effluent and Environmental Radiation Surveillance, J. Kelly, (Ed.), 63 74American Society for Testing and Materials, Johnson, USA

23. C. Kim, M. Han, 1999Dose assessment and behavior or tritium in environmental samples around Wolsong nuclear power plant. Appl Radiat Isot, 50 783 791

24. C. Liu, J. Chao, C. Lin, 2003Tritium release from nuclear power plants in Taiwan. Health Phys, 84 361 367

25. G. Lopez-Abente, et al. 1999Leukemia, lymphomas, and myeloma mortality in the vicinity of nuclear power plants and nuclear fuel facilities in Spain. Cancer Epidemiology, Biomarkers & Prevention, 8 925 934

26. F. Luykx, G. Fraser, 1986Tritium releases from nuclear power plants and nuclear fuel processing plants. Radiat Prot Dosim 16 31 36

27. NCRP 1985Carbon-14 in the environment- NCRP Report 81NCRP, ISBN, Bethesda, USA

28. NCRP 1987Public radiation exposure from nuclear power generation in the United States- NCRP Report 92NCRP, ISBN, Bethesda, USA

29. T. Nedveckaite, et al. 2000Environmental releases of radioactivity and the incidence of thyroid disease at the Ignalina NPP. Health Phys, 79 666 674

30. C. Patrick, 1977Trends in public health in the population near nuclear facilities: a critical assessment. Nucl Saf, 18 647 662

31. Poncela. L. Quindos, et al. 2003Natural radiation exposure in the vicinity of Spanish nuclear power stations. Health Phys, 85 594 598

32. B. Shleien, A. Ruttenber, M. Sage, 1991Epidemiologic studies of cancer in populations near nuclear facilities. Health Phys, 61 699 713

33. H. Tokuyama, M. Oonishi, 1997Precipitation washout of tritiated water vapor from a nuclear reactor. J Environ Radioact, 34 59 68

34. United Nations Scientific Committee on the Effects of Atomic Radiation 2000 Sources and Effects of Ionizing Radiation, UN, 978-9-21142-239-9New York, USA

35. United States Census Bureau (2010). Statistical abstract of the United States: 2010, GPO, Washington, D.C., USA

36. USEPA (1977). Environmental radiation protection standards for nuclear power operations. 40 CFR 190, 42 FR 2860, January 13, 1977, GPO, Washington, D.C., USA

37. USNRC 1975 Programs for monitoring radioactivity in the environs of nuclear power plants. Regulatory Guide 4.1, rev. 2, GPO, Washington, D.C., USA

38. USNRC 1976a Calculation of releases of radioactive materials in gaseous and liquid effluents from boiling water reactors, NUREG-0016, GPO, Washington, D.C., USA

39. USNRC 1976b Calculation of releases of radioactive materials in gaseous and liquid effluents from pressurized water reactors, NUREG-0017, GPO, Washington, D.C., USA

40. USNRC 1978 Preparation of radiological effluent technical specifications for nuclear power plants, NUREG-0133, GPO, Washington, D.C., USA

41. USNRC (1991). Standards for protection against radiation. 10 CFR 20, 56 FR 23361, May 21, 1991, GPO, Washington, D.C., USA

42. USNRC (1995). Dose commitments due to radioactive releases from nuclear power plant sites in 1991, NUREG/CR-2850, GPO, Washington, D.C., USA

43. USNRC (1996). Technical specifications on effluents from nuclear power reactors. 10 CFR 50.36a, 61 FR 39299, July 29, 1996, GPO, Washington, D.C., USA

44. E. Vold, 1983Ingestion pathway factor in dose assessment for annual airborne releases of radioactivity. Health Phys, 47 429 441

45. A. Walmsley, et al. 1991The distribution of doses to members of the public around UK civil nuclear sites. Radia Protect Dosimetry, 36 215 218

46. P. Ziqiang, et al. 1996Radiological environmental impact of the nuclear industry in China. Health Phys, 71 847 862

Chapter 5

Power System Protection Design for NPP

Chang-Hsing Lee[1] and Shi-Lin Chen[2]

[1] EE Dep. of National Tsing Hua University, Hsinchu, Taiwan

[2] EE Dep. of Chung Yuan Christian University, Chung Li, Taiwan

1. INTRODUCTION

One of the key purposes of NPP power system protection is to ensure that NPP's local power demand (such as cooling pumps, control systems, etc.) are met under all circumstances even during faulted periods. To achieve this goal, NPP power system protection must ensure that it can supply these local loads using either (1) power from the grid (via the transmission connection, which in most time, however, are used for exporting the excess power generated by the NPP after supplying its local loads) or (2) power from local generations such as diesel generators, batteries, etc. at all times and under all circumstances.

On the first power source (grid power), many NPPs worldwide have been built along the seashore for cooling water availability reasons. Overhead transmission lines are thus built in the vicinity of the seashore to transport the large amount of power generated from the NPP to the grid economically. As these overhead lines are exposed to salt contamination, flashover will occur when contamination becomes excessive. In the event of flashover, which is equivalent to a line-to-ground fault, the plant's protection system will need to initiate a series of switching operation to redirect the large power output from the NPP to a backup route in order to avoid reactor emergency shut-down. However, such switching has the adverse effect of causing undesirable transient overvoltages to propagate in the plant's local power grid [1-4]. Dealing with the frequent switching actions of these overhead lines while mitigating their adverse effects thus becomes the first challenge of designing NPP power system protection.

Once the NPP loses its connection to the grid, it will need to rely on the local generation to continue supplying its local loads. Most NPP use multiple "independent" sources as backup power. However, unless NPP's local power grid is properly configured and its protection system properly designed, these "independent" sources can all fail at the same time as manifested in Taipower's 18 March, 2001 Level 2 event ("318 Event") [5].

In the following sections, we will examine Taipower's "318 Event" in detail to demonstrate the various possibilities that could lead to NPP plant blackout. Moreover, as these possibilities are not mutually exclusive, we will use this example to illustrate how multiple or cascaded problem can present further challenges to the overall NPP power system protection design. Recommended preventive measures are then summarized in the final section of this chapter.

2. TAIPOWER "318 EVENT"

2.1. System Configuration

Figure 1 shows the configuration of Taipower 3rd nuclear power plant. The NPP has two 951 MW generators which are connected to the local 345kV gas-insulated substation (GIS) in one-and-half breaker configuration as shown in Fig. 1. The NPP is then connected to the power grid via four 345 kV overhead power lines (Darpen 1, 2 and Lunchi Sea/Mountain) to the Darpen and Lunchi 345kV EHV (Extra High Voltage) substation and two 161kV overhead lines (Kengting and Fengkang) to Kenging and Fenkang 161kV HV (High Voltage) substation.

It is important to note that there are three 13.8 kV buses (in the middle) and four 4.16 kV buses (at the bottom) for plant utility. Among these buses, the two 4.16 kV buses in the middle are responsible for feeding the safety-critical equipment such as cooling pumps and are designated as "essential buses". (The 2nd 4.16kV bus from the left where "DGA (Diesel Generator A)" is connected is designated as "Essential Bus A". The one next to it, where "DGB" is connected to, is "Essential Bus B".)

Another notable but subtle feature of this configuration is the use of 3-phase gas-insulated line (GIL) design with the 3 phases enclosed in a single duct of approximately 340 meters for the connection of the generation units and auxiliary systems, (located at the foot of a hill) to the 345kV GIS (on the top of the hill) due to topography feature of the location. This feature has implication on the generation and propagation of switching transients which will be explained in later sections.

2.2. Event Sequence

On 18 March, 2001, a Level 2 event occurred at Taipower's 3rd NPP, and the whole plant went into blackout from 00:45 to 02:58. The event started at 00:45 when EHV CB3510 (see Fig. 1, highlighted in red) was closed to energize the then-offline 345 kV/13.8 kV/4.16 kV start-up transformer (X01). Upon CB3510 closure, medium voltage (MV) CB#17 on "Essential Bus A" exploded damaging not only CB#17 but also CB#15. CB#15 formed a permanent ground fault keeping "Essential Bus A" at ground potential thus Essential Bus A became useless. CB #3 and #5 on "Essential Bus B" was then opened hoping DGB will start and supply power to those critical loads. However, DGB failed to start and the whole plant went into blackout. The only hope remained at that time was DG5 (on the far right in Fig. 1) which, however, needs to be started locally and manually. After 2 hour since the first problem occurred, DG5 was finally started and started to supply power to the critical loads via "Essential Bus B".

Figure 2 shows the switching event log of the four 345kV and two 161kV lines connecting to the NPP. After reviewing the event log, it was found that CB#17 on Essential Bus A broke down when one GIS switching operation was occurring. The event log also showed that there were 37 EHV switching operations during the 48-hour period prior to the event due to salt-fog influence in the plant area. Because of the unstable offsite power, the GIS switched between different offsite power to acquire the stable power sources.

Figure 3 shows the transient recording of overvoltages for both the 345kV bus and one MV (medium voltage) bus at 20:38 in March, 2001. At t0, the flashover on the 345kV line occurred leading to its subsequent tripping at t1. The tripping took place on the remote end of the 345kV line thus overvoltage can still be observed at the NPP between t1 and t2 due to "motor-generating effects" to be explained in the following section of this Chapter. The overvoltage on the 345kV line eventually caused flashover from Phases A and B to ground pulling down the line voltage and all motors on the 4.16kV bus were tripped by their respective under voltage relays at t3. At t4, the flashover from Phases A and B to ground was cleared and the "motor-generating effects" start to build up the voltages again with the two remaining motors on the 13.8kV bus.

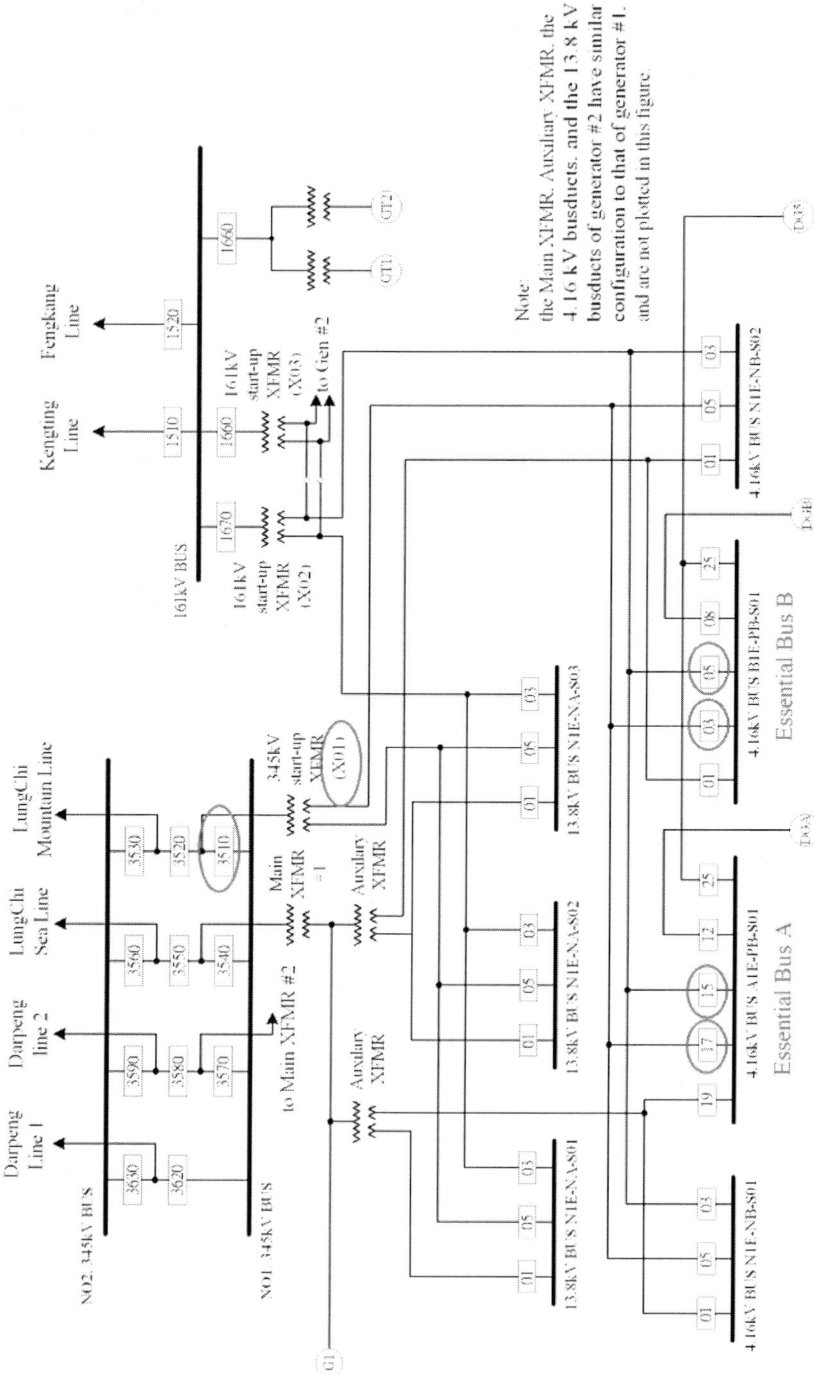

Figure 1. System Configuration of the NPP

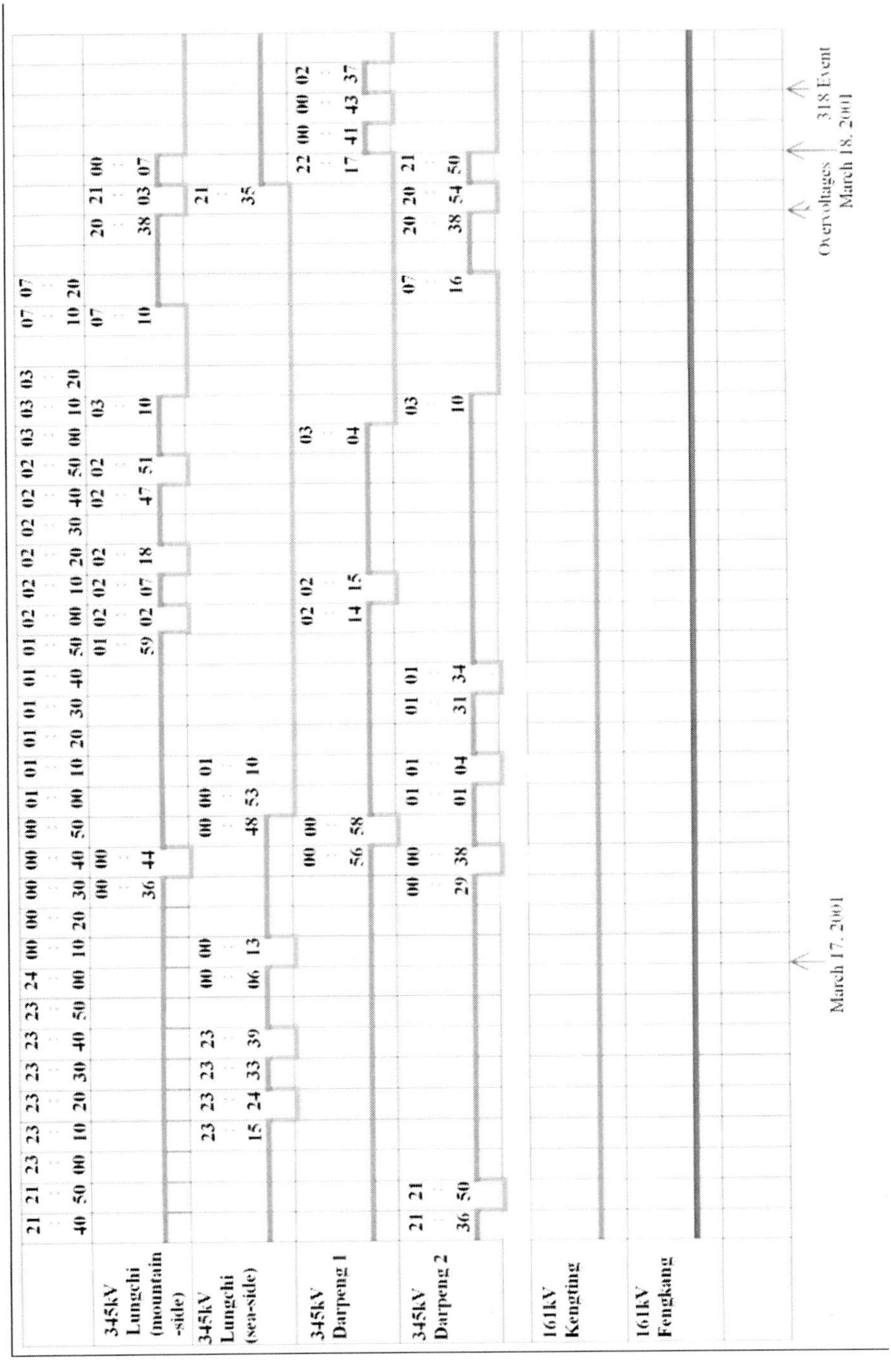

Figure 2. 345kV and 161kV Overhead Line Switching Event Log

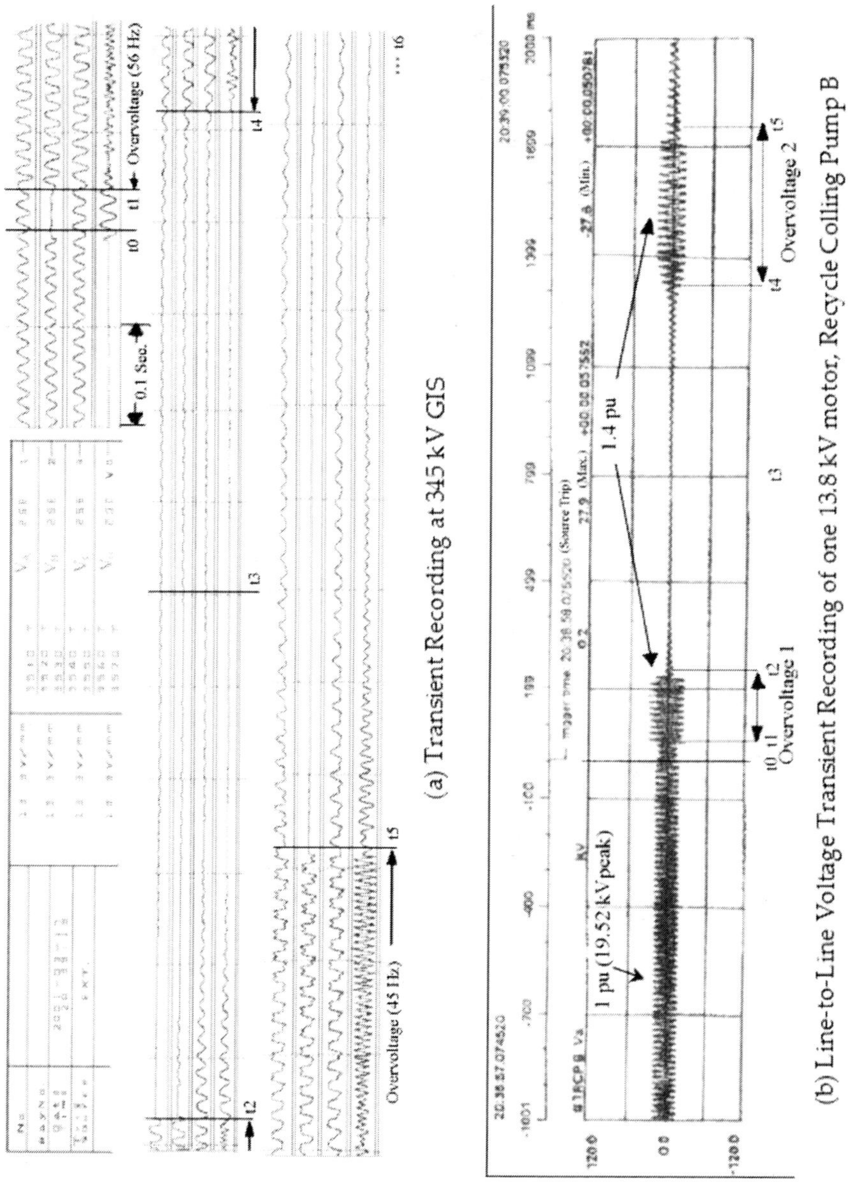

Figure 3. Overvoltages at 20:38 in March 17, 2001

2.3. Electrical Stress in Plant Power System

2.3.1. Line Conductor Overvoltages Due To Over-Excitation And Nonlinear Resonance [6,7]

The transient recorder in Fig. 3(a) recorded 2 abnormal overvoltages (at 56 Hz and 45 Hz, respectively) after the last 345 kV-transmission line connecting to the NPP was tripped on the remote end which turned the NPP into an electrical island. As will be explained in the next section, the 1st overvoltage was caused by the over-excitation of the motors (e.g. recycle water pump) in the plant who, with terminal voltages supported by large line capacitance, now operated as induction generator after loss of external power.

The 2nd overvoltage is caused by a different mechanism. After a few cycles the low voltage relays tripped many of the plant motors leaving only 2 biggest motor (now operating as induction generator) still connected and were supported by a comparatively much larger capacitance leading to not only over-excitation but also magnetic saturation of both the motors and transformers. This created a condition very close to ferroresonance resulting in even bigger overvoltage.

2.3.2. Neutral Voltage Transfer

It can be seen from Figure 3 that overvoltage were observed not only on the line conductors of phase A, B, and C but also on the neutral. As will be explained in the next section, neutral voltage transfer can occure through 2 different mechanisms: electromagnetic and capacitive transfer.

In the presence of transformer core saturation, 3rd harmonic neutral voltage will be present on the windings through electromagnetic transfer as long as the neutrals of the respective windings are not grounded. In the presence of neutral voltage on any of the transformer windings, the stray capacitance among the windings and earth will result in capacitive neutral voltage transfer.

2.3.3. Switching Surges on Both Ehv And Mv Systems

From Fig. 2, it can be seen that there were 37 switching operations during the 48-hour period prior to the breakdown. This unusually high number of switching operation can create lots of switching surges (with magnitude of around 7 times the rated line-to-ground peak voltage in the medium voltage system) which, when propagating through the NPP local power network, can degrade the insulation level or even cause breakdown of CB's in the local power network [8,9].

It should be noted that while there were 37 switching operations on the EHV side, none of the switching surges were captured by the transient recorder in the 345kV GIS in Fig. 3(a) due to insufficient bandwidth of the transient recorder. In a follow-up field test [9] after the event, it was found that such switching often causes switching surges of around 7 times the rated line-to-ground peak voltage!

3. STRESS MECHANISM AND MODELING

It can be seen from the above that Taipower's 3rd NPP was under sigificant and multiple stresses before and during the Level 2 event. This section explains the mechanisms working behind these stresses and provide basic principles how to model them.

3.1. Line Conductor Overvoltages Due To Over-Excitation And Nonlinear Resonance [6,7]

Figure 3 shows that on Phases A, B, and C there were two overvoltages observed where the second overvoltage was slightly higher than the first. Causes of these 2 overvoltages are detailed as following.

3.1.1. First Overvoltage (56hz) – Over Excitation

Figure 4 shows the 2 essential condition for induction motor generating effect: large capacitance and continuous rotating motor. When an induction motor lost its external voltage source, the flywheel with large inertia will keep the motor rotating and the capacitance of transmission line will provide the necessary voltage support for the induction motor to act as a generator. The magnitization curve of the motor and the amount of capacitance will jointly determine the overall motor generating effect as shown in Fig. 5. If the capacitance is too small to provide enough magnetizing current (curve C0 in Fig. 5), the terminal voltage of motor will decay exponentially and the generating effect will not sustain. However, if the capacitance is large enough, the motor generating effect will sustain and the terminal voltage is determined by the intersection of the capacitance and magnetizing curve such as (V1, C1) and (V2, C2) in Fig. 5.

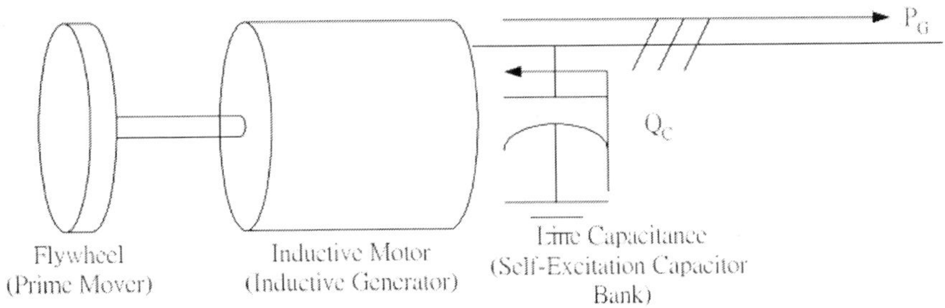

Figure 4. Equivalent Circuit of Motor-Generating Effect.

At "t1" in Fig. 3(a), the last 345kV transmission line connecting to the NPP was tripped on remote end due to a flashover on the line turning the NPP into an electrical island. As the local end of the 345kV line did not trip, a motor generating condition equivalent to Fig. 4 was formed with the 127kM transmission line providing sufficient capacitance to support the voltage of the various motors in the NPP. As can be seen in Fig. 3(b), the terminal voltage is increased to 1.4 p.u. and the overall resultant frequency is 56 Hz.

During this first overvoltage period, the terminal voltage of motor was about 1.4 p.u. (Fig. 3(b)) but the line voltage was about 1.29 times the rated line-to-ground peak voltage (Fig. 3(a)). This implied that the power transformers have saturated. As a result, a lot of harmonics were produced and the zero sequence components of them would be integrated into the neutral voltage resulting in unexpected high neutral voltage. This period ended at "t2" in Fig. 3(a) when the flashover grounded both phase A and B.

Table 1 shows the harmonic contents of B phase voltage between t1 and t2 in Fig. 3(a). The even order harmonics and DC component could be treated as the slight magnetic bias caused by asymmetric fault. At this stage, there was no ferromagnetic resonance in the island system.

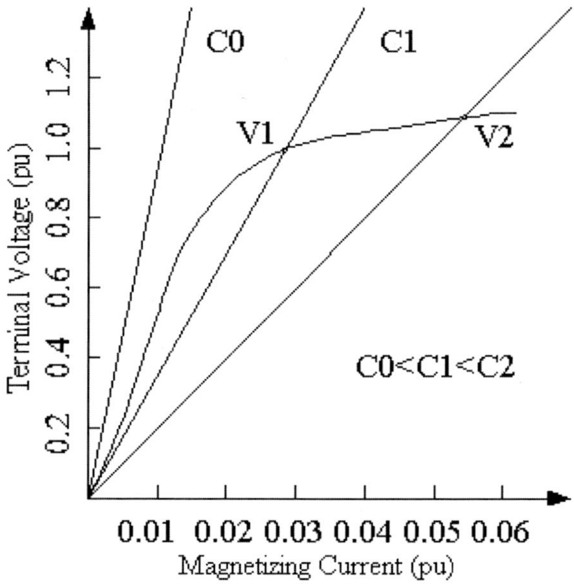

Figure 5. Relationships between Motor Terminal Voltage, Magnetization Curve, and External Capacitance

Table 1. Voltage Harmonic Contents of Phase B between t1 and t2

Order	DC	1	2	3	4	5	6	7	8
%	9.3	100	7.8	8.0	3.6	13.5	1.6	6.1	1.2

3.1.2. Second Overvoltage (45 Hz) - Nonlinear Resonance

Figure 6 shows the four essential conditions for a ferroresonance to occur: voltage source, capacitance, nonlinear inductance (ferromagnetic and saturable), and low losses. The R in the RLC resonant circuit in Fig. 6 is very large due to the "low losses" condition and can often be ignored. The nonlinear inductance L is the magnetizing curve of the motors and transformers in the system and the capacitance is provided by the transmission line.

At "t3" in Fig. 3(a), all the motors on the 4.16 kV system were tripped by undervoltage relay. Between t3 and t4, the flashover grounding of phases A and B were cleared and the motor generating effects mentioned above picked up again gradually re-establishing the line voltage. At "t4" in Fig. 3(a), most motors in 13.8 kV system were also tripped by undervoltage relay with the exception of two largest ones. With the capacitance provided by the transmission line now need only to support the terminal voltage of 2 motors, we would expect the terminal voltages to be higher than those during the first overvoltage stage according to Fig. 5. However, due to deep saturation of the motors and transformers, the overvoltage magnitude in Fig. 3(b) during the 2nd overvoltage is only slightly higher than the previous stage. This can be further seen from the fact that at the beginning of “t4” in Fig. 3(a), there were no overvoltage and no distortion of waveforms. As line voltage increased, the harmonics increased and after a few cycles the amplitude of voltage remained but voltage waveform distorted dramatically. Figure 7 shows the waveform at 4 cycle prior to t5 with its Fourier components summarized in Table 2 [10].

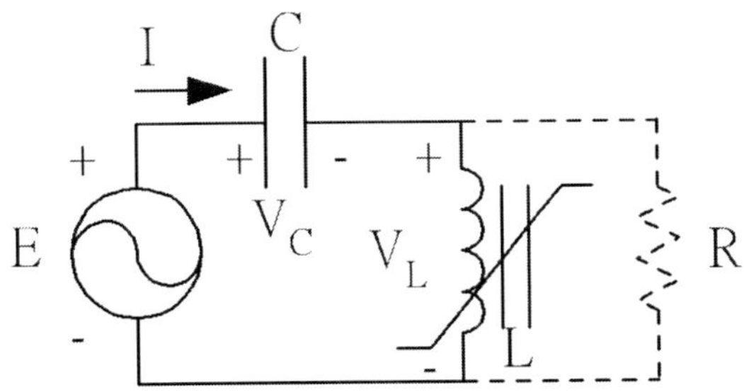

Figure 6. Equivalent RLC Circuit

It can be seen from Table 2 that the voltage of fundamental frequency was about 1.5 times the rated line-to-ground peak voltage, which is slightly larger than the previous overvoltage. The large DC and even-order harmonics indicate the deep saturation of the start-up transformer. In Table 2 the total of 3rd harmonics is 554.7 kV_{peak} (Note: The 3rd harmonics are in-phase therefore can be added up directly.) Comparing this figure with the neutral voltage of 626 kV_{peak} in Fig. 3(a), this indicates that 3rd harmonics is the main source of neutral voltage during the second overvoltage period.

As the inductances in the systems are now deeply saturated, there is a possibility that ferroresonance can occur. (Note: Ferroresonance is nonlinear resonances in power system where the voltage and current may change from normal steady state to another steady state with large harmonic distortion.) The phenomenon can be best understood from a circuit perspective using Figure 6 as example. In Fig. 6, the total equivalent impedance of the circuit is (jX_L - jX_C). When the inductance is saturated and current further increases, it will drive the inductor into deeper saturation where the inductor impedance jX_L will reduce when current further increases. A critical point will be reached at Point B in Figure 8when (jX_L - jX_C) becomes zero. Any current increase beyond Point B will cause the total impedance change from a positive value to a negative value causing resonance effects near this operating point[7].

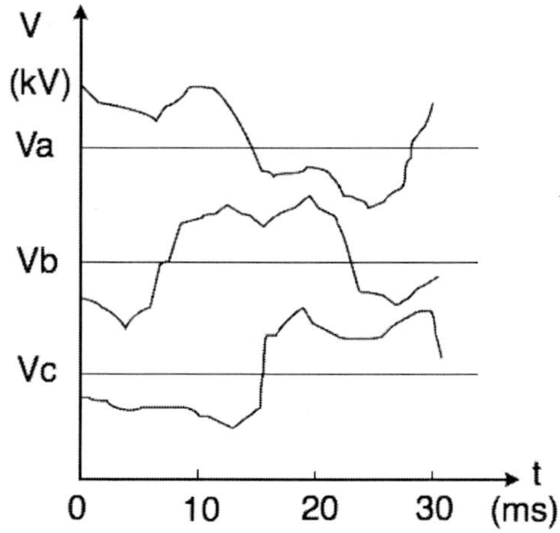

Figure 7. Zoom-in of The 4 Cycles prior to t5

Table 2. Fourier Analysis of **Fig. 7**

order	Phase A		Phase B		Phase C	
	V_{peak}	$\%_{fund.}$	V_{peak}	$\%_{fund.}$	V_{peak}	$\%_{fund.}$
0	175.068	0.429	48.279	0.113	71.93	0.017
1	408.273	1.0	427.991	1.0	413.726	1.0
2	108.386	0.265	47.287	0.11	39.829	0.096
3	179.679	0.44	179.734	0.42	195.357	0.472
4	41.29	0.101	28.313	0.066	24.702	0.06
5	28.13	0.069	30.829	0.072	65.626	0.159
6	13.036	0.044	10.364	0.024	22.34	0.054
7	12.562	0.031	22.126	0.052	37.579	0.091
8	15.84	0.039	28.219	0.066	7.708	0.019
9	14.315	0.035	8.207	0.019	39.509	0.095
10	9.428	0.023	16.289	0.038	6.203	0.015
11	18.092	0.044	14.962	0.035	27.434	0.066
12	18.805	0.046	27.966	0.065	25.323	0.061
13	15.387	0.038	11.466	0.027	22.233	0.054
14	15.717	0.038	10.323	0.027	24.685	0.06
15	13.897	0.037	24.991	0.058	13.998	0.034
16	1.992	0	36.961	0.086	12.173	0.029

Based on analysis of all available data, it is believed that the 2nd overvoltage from t4 onward is on the boundary to be ferroresonance therefore the 2nd overvoltage is caused by a combination of motor-generating effect and nonlinear resonance.

3.2. Neutral Voltage Transfer [10]

It can be seen from Figure 3 that overvoltage can be observed not only on the line conductor but also on the neutral conductor as well. In order to understand this phenomenon we need to look at Fig. 9 where the equivalent circuit of a transformer is shown including its stray capacitances.

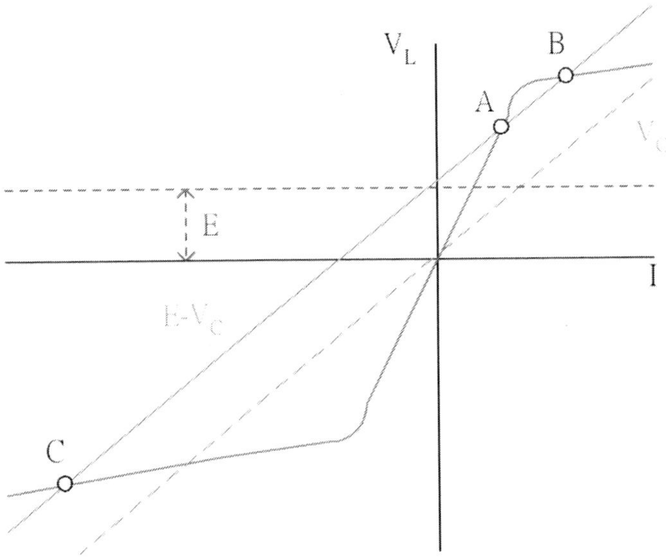

Figure 8. Ferroresonance Phenomenon Explanation

3.2.1. Transformer Modeling

In Fig. 9, C_{HE} and C_{LE} depict the stray capacitance between high voltage (HV) winding to ground, and low voltage (LV) winding to ground, respectively, while C_{HL} depicts the stray capacitance between HV and LV windings. Typical stray capacitances for the 345/13.8/4.16kV power transformer are shown in Table 3.

In the presence of transformer core saturation, 3rd harmonic neutral voltage will be present on the windings through electromagnetic transfer as long as the neutrals of the respective windings are not grounded. Once the neutral voltage is established on any side of the neutrals, the stray capacitance provides a further path for it to transfer to other neutrals according to Equation (1)

$$E_{L0} / E_{H0} = C_{HL} / (C_{HL} + C_{LE}) \tag{1}$$

where E_{H0} is the neutral voltage at HV side, and E_{L0} is the neutral voltage at LV side.

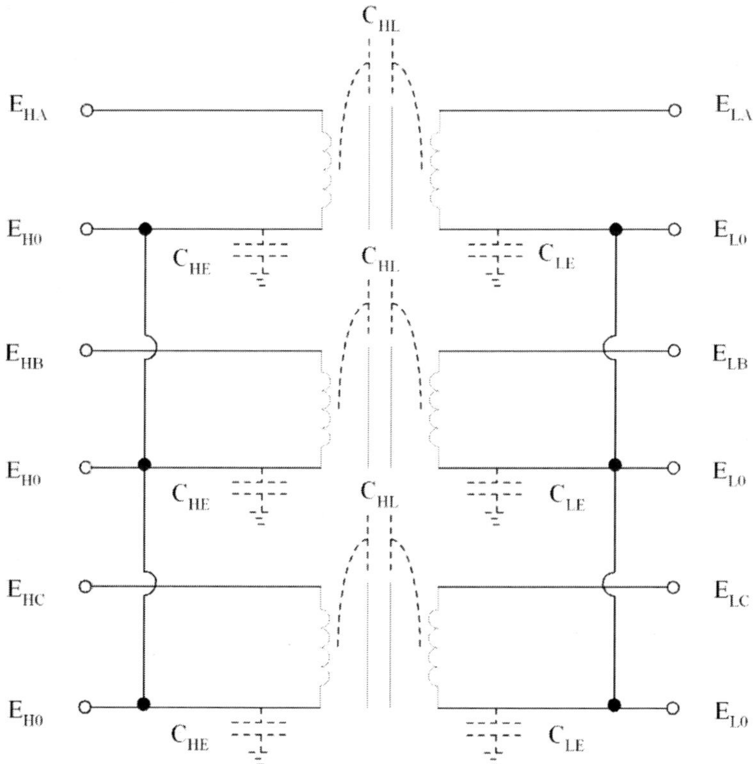

Figure 9. Voltage Transfer Diagram of 345 kV/4.16 kV Transformer

3.2.2. Neutral Voltage Transfer

It can be seen from Fig. 3(a) that, during 1st overvoltage the neutral voltage gradually roses to 200 kV_{rms} while during 2nd overvoltage the neutral voltage rose to 626 kV_{peak} (Note: the voltage waveform became very non-sinusoidal during 2nd overvoltage, we thus use peak value instead of rms value). The source of both overvoltages in the neutral was due to motor and transformer saturation resulting in 3rd harmonic voltages at the neutral however during the 2nd overvoltage the waveform is much more distorted with higher harmonic content.

Table 3. Stray Capacitance of the 345kV/13.8kV/4.16kV Power Transformer

Item	$C_{345/Earth}$	$C_{13.8/Earth}$	$C_{4.16/Earth}$	$C_{345/13.8}$	$C_{13.8/4.16}$	$C_{345/4.16}$
Capacitance	4.48 nF	13.76 nF	21.92 nF	4.3 nF	214.86 pF	8.96 nF

As indicated by Fig. 3(a), the neutral on 345kV side does not appear to have been effective grounded possibly due to grounding failure. The result is that very high neutral voltage was established on the 345kV neutral and if the 4.16kV neutral was not grounded it will see a neutral voltage (through capacitive neutral transfer) of

$$200kVrms * \frac{8.96nF}{8.96nF + 21.92nF} = 58.03kVrms$$

$$626kVpeak * \frac{8.96nF}{8.96nF + 21.92nF} = 181.54kVpeak$$

during 1st and 2nd overvoltages, respectively. To better appreciate various grounding combination's effect on the neutral voltage transfer, 3 simulations were conducted assuming grounding conditions as per Table 4 and their results are summarized in Table 5.

Table 4. Grounding Condition for Simulating Capacitive Transfer

Item	345kV side	13.8kV side	4.16kV side
Case 1	Ground (direct)	Ground (8 Ω)	Ground (2.4 Ω)
Case 2	Non-Ground	Non-Ground	Non-Ground
Case 3	Non-Ground	Ground (8 Ω)	Ground (2.4 Ω)

Table 5 shows that as the neutral voltages transferred to the 4.16kV bus can be as high as 13 times the phase-to-ground peak voltage which can pose significant threat to CB#17 as well as other CB's. However, if the neutral systems were properly configured, the risk can be minimized greatly.

Table 5. Capacitive Transfer Simulation Result

	345kV side Phase Voltage	345kV side Neutral Voltage	4.16kV side Neutral Voltage
Case 1	457.55 kV_{peak}	0 kV_{peak}	0 kV_{peak}
Case 2	450.6 kV_{peak}	275.88 kV_{peak}	42.29 kV_{peak}
Case 3	448.93 kV_{peak}	44.58 kV_{peak}	0.002 kV_{peak}

(a) 1st overvoltage

	345kV side Phase Voltage	345kV side Neutral Voltage	4.16kV side Neutral Voltage
Case 1	428.83 kV_{peak}	0 kV_{peak}	0 kV_{peak}
Case 2	418.7 kV_{peak}	263.43 kV_{peak}	41.08 kV_{peak}
Case 3	420.94 kV_{peak}	46.25 kV_{peak}	0.002 kV_{peak}

(b) 2nd overvoltage

3.3. Switching Surges and Very Fast Transient Overvoltage (Vfto)

Switching operations are the most prominent phenomenon in the "318 Event". During the 48 hours prior to the Level 2 event, there were 37 switching operations and each could cause switching surges. Switching surges caused by GIS switching is characterized by its nanosecond wavefront and is commonly referred to as Very Fast Transient Overvoltage (VFTO) [11].

VFTO is the phenomenon of transient overvoltage generated during switching operation characterized by very short rise-time of 4 to 100 ns and has been covered by various literatures [3,12-22]. The phenomenon is particularly significant during Disconnect Switch (DS) operation due to multiple-restriking in the DS due to lack of arc-suppressing chamber.

3.3.1. Field Measurement

In the past, VFTO was not considered to be possible to transfer from EHV through power transformer to medium voltage (MV) system [12-14]. However, in light of the "318 Event", a field test was conducted in Taipower 3rd NPP during plant overhaul by switching the DS of EHV GIS and measure the voltage on "Essential Bus A".

Field test result [9] shows that after switching the DS of EHV GIS, multiple 25 kV-level restrikes (approximately 7 times the rated line-to-ground peak voltage) were measured on the 4.16kV bus indicating VFTO can be transferred from the EHV side to MV side. It also indicates that the maximum peak voltages measured on the 4.16kV bus occur neither on the first strike nor on the last strike, and this behaviour is quite different with that in EHV system. The measurement results are shown in Fig. 10.

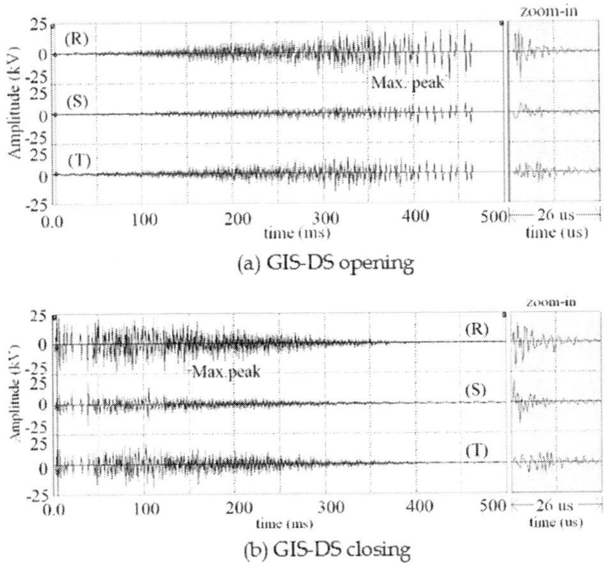

Figure 10. Switching Surge Measured on 4.16kV Bus by Operating EHV GIS Disconnect-Switch. (Note: The bandwidth of the measurement system was 2MS/s, the highest achievable in 2003)

3.3.2. Vfto Simulation

To further appreciate VFTO transfer mechanism, numerical simulation model was built [23]. To validate this simulation model, the field test condition for Fig. 10 was reconstructed and the simulation result is shown in Fig. 11. It can be seen from Fig. 11 that the waveform envelope are consistent with measurement for both DS opening and closing and that the maximum VFTO on the essential bus occurred neither at first nor at last strike.

We then change the DS operation angle for each 5° intervals to simulate different closing/opening condition and Table 6 and 7 summarizes the maximum EHV inter-contact breakdown voltage vs. maximum MV VFTO. The following can be observed from Table 6 and 7:

1 The VFTO transferred to the essential bus A can be as high as 28.77 kV, which is about 8.47 times the rated line-to-ground peak voltage.

2 For all simulations, the restrike that causes the maximum VFTO on "Essential Bus A" does not necessarily coincide with the one that caused the max inter-contact breakdown voltage on EHV side.

3 Among the 36 simulations for DS opening, the simulation that produces the highest inter-contact breakdown voltage on EHV side is not the same as the one that produces the maximum VFTO on "Essential Bus A". This is also true for DS closing. E.g., Case #28 (δ_{oper} = 135°) of DS opening produces the highest VFTO in MV system (28.77 kV) while it was Case #18 (δ_{oper}=85°) that produces the highest inter-contact breakdown voltage on EHV side (354.2 kV).

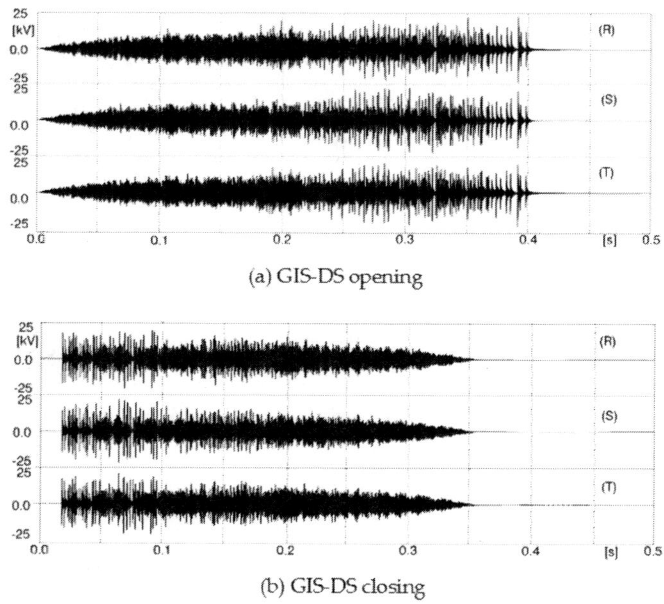

Figure 11. Simulation of VFTO at the Field Measurement Point

Table 6. Max. Inter-contact Breakdown Voltages vs. Max. VFTO in MV for DS Opening

Item	Among the Multiple Restrikes				Total Num. of Restrikes on EHV Side per Φ
	Max. Inter-contact Breakdown Voltage		Max. VFTO at 4.16 kV		
	Mag. (kV)	Seq. Num.	Mag. (kV)	Seq. Num.	
Max. Inter-contact Breakdown Voltage Case # 18 among 36	354.2	467	23.76	444	468
Max. VFTO at 4.16 kV Bus Case #28 among 36	323.6	447	28.77	432	447

Table 7. Max. Inter-contact Breakdown Voltages vs. Max. VFTO in MV for DS Closing

Item	Among the Multiple Restrikes				Total Num. of Restrikes on EHV Side per Φ
	Max. Inter-contact Breakdown Voltage		Max. VFTO at 4.16 kV		
	Mag. (kV)	Seq. Num.	Mag. (kV)	Seq. Num.	
Max. Inter-contact Breakdown Voltage Case #14 among 36	284.8	1	23.27	12	460
Max. VFTO at 4.16 kV Bus Case # 23 among 36	278.0	1	26.41	26	455

3.3.3. Characteristic of Vfto Transferring To Mv System

3.3.3.1. Capacitive Coupling Of High-Turn-Ratio Transformer

VFTO and the oscillation voltages V_{OSC} (voltages created by preceding restriking that can be superimposed to the following restrike) on the EHV side can be transferred to MV system through the start-up power transformer via capacitive coupling. The transfer ratio is mainly dependent on transformer's EHV-to-MV interwinding capacitance, transformer's MV winding-to-enclosure capacitance, and the bus-to-ground capacitance of MV system [23]. From both our measurement and simulation result, it was observed that the V_{OSC}, which is of several tens kV in the EHV GIS, could still be of several kV in the MV system, and this will be superimposed to the VFTO coupled from the EHV side causing up to 7 ~ 8.47 times the rated line-to-ground peak voltage on MV side.

3.3.3.2. Superposition of oscillations initiated by a prior strike on top of subsequent restrikes

Figure 12(a) shows two consecutive restrikes from a multiple-restrike simulation and Fig. 12(b) shows its counterpart single-strike simulation. It can be seen from Fig. 12(a) that the V_{OSC} initiated by the first restrike is superimposed to the second restrike resulting in a higher peak voltage (10.72 kV vs. the single strike one of 9.88kV).

3.3.3.3. Maximum VFTO transferred to MV for DS closing vs. DS opening

During DS opening the contact distance becomes wider and wider leading to longer intervals between two consecutive restrikes while that during DS closing is the opposite. As a result, there is a higher probability of superposition of V_{OSC} to subsequent restrike during DS closing (thus higher VFTO) than opening.

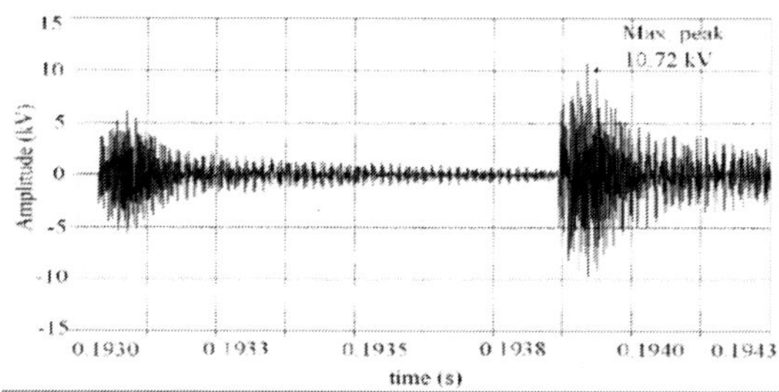

(a) Two consecutive restrikes taken out of multiple-restrike simulation of DS opening

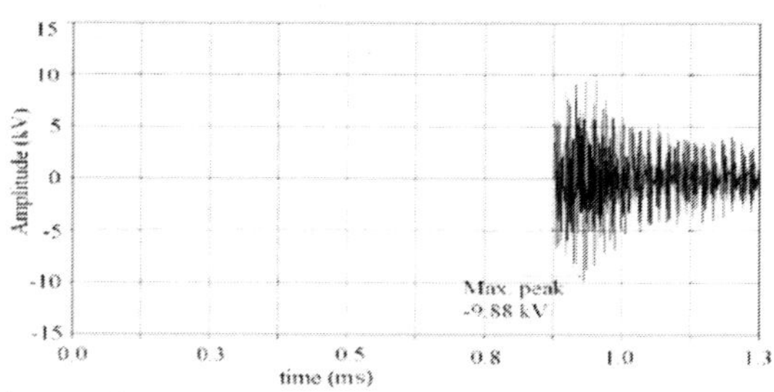

(b) Single-strike simulation with the same inter-contact breakdown voltage as in (a)

Figure 12. Oscillation Voltage (V_{OSC}) Initiated by a Strike or Restrike Can Be Superimposed to Subsequent Restrike Voltages.

4. LESSON LEARNED AND IMPORTANT ASPECTS OF NPP POWER SYSTEM PROTECTION DESIGN

Events like the "318 Event" were seldom caused by one single reason. It can be seen from the above discussion that Taipower 3rd NPP was under multiple stresses before the event and there were mutiple mechanisms for the generation, amplification, and transfering of overvoltages which, combined with the operation practices and equipment history, eventually led to the explosion of CB#17 and total blackout of the NPP. Below are the key lessons learned from this event and their recommended preventive measure.

4.1. Bus Configuration and Fault Area Isolation

The "318 Event" was essentially triggered by a single equipment failure but leading to a complete blackout of the power plant. There are 2 key lessons learned from this event: (1) Explosion of CB#17 took down the adjacent CB#15 as well. (2) "Independent sources" are not always independent due to improper bus configuration.

For various reason such as space requirement, ease of maintenance, etc, switchgear panels are usually installed in the same room side by side. If this cannot be changed, during the risk evaluation process one must consider the N-1 condition being loss of "one group of equipment" instead of "one equipment" unless sufficient separation are provided between the equipments.

The "independence" of power sources need then be examined closely. If multiple sources or multiple buses can be taken down by a single failure such as permenant fault to ground, etc, they cannot be considered as independent sources and more backup needs to be added.

It should be noted that during the "318 Event", after the explosion of CB#17 the plant utility room was filled with smoke which makes the manual starting of other diesel generators extremely difficult. Not only were equipments under significant stress but also the human operators. It is thus recommended that the feasiblity of starting backup sources under utility room smoke condition be checked and that any manual operation required during this stage be as simple and straightforward as possible with proper interlock to reduce the chance of human error which may further escalate the event.

4.2. Nonlinear Resonance Prevention

Among all the scenarios considered in this Chapter, nonlinear resonance is the most difficult one to be detected. In view of the potential hazard it could cause, precautionary measure must be taken to prevent it from initiating.

The first step is to prevent motor-generating effect from ever occuring (thus removing the key source of initiation.) As explained above, the essential conditions of motor-generating effect are (1) rotating motor with large inertia, (2) large capacitor bank in an isolation system to support the terminal voltage. Since a rotating motor with large inertia can not be stop immediately, the focus is to remove the capacitive support. In the case of Taipower 3rd NPP, the capacitive support came from the long transmission line who were tripped only on the remote end. It is recommended that Direct Transfer Trip (DTT) function be implemented for transmission line protection to greatly reduce the risk of motor generating effect.

The second step is to ensure effective grounding of transformer neutrals as designed. Due to the objective of minimizing short circuit current, the neutral groundings in NPP are usually multi-configured: arrestor grounded under normal condition and direct grounding when in islanding operation. The switching from one grounding scheme to another often requires manual operation and this increases the risks of leaving the islanded system ungrounded as well as nonlinear resonance of power. Proper interlock or checking mechanism should be implemented to ensure proper grounding as designed at all times.

4.3. Neutral Voltage Transfer

Neutral voltage transfer can occur via either electromagnetic or capacitive transfer. Based on simulation result the risk can be significantly reduced with proper grounding of the neutral. This, however, must be carefully implemented in order not to increase the short circuit current in the NPP.

Again, any manual operation during event would introduce extra risks therefore should be desigined to be as simple and straightforward as possible with proper interlock or checking system.

4.4. Vfto Transferring To Mv System

According to field measurement and numerical simulation, the VFTO transferring to MV system is usually underestimated by literatures. As demonstrated by both the field measurement and simulation result, peak voltage of VFTO in MV system could be as high as 8.47 times the rated line-to-ground peak voltage with an average 466 times restrike during DS operating [1,23]. Though the peak VFTO voltages transferred to the MV side are usually still within the basic impulse insulation level (BIL) tolerances of the equipment, this does not mean that repeatedly striking the equipment with 8.47 times the rated line-to-ground peak voltage would cause no damages to the equipment. In fact, this can accelerate equipment ageing and cause quick degradation of the insulation material and eventually leading to equipment breakdown.

After the "318 Event", a recommendation was made to Taipower No. 3 Nuclear Power Plant in 2003 for the installation of surge absorbers (0.8μF capacitor specially designed for surge absorption installed right close to the start-up transformer for each of the three phases) on the MV side in Fig. 1 [9,23]. The recommendation was adopted by Taipower in March 2005 and a subsequent measurement in March 2006 plus one-year monitoring indicated that there were no further VFTO exceeding rated line-to-ground peak voltage on the MV system.

4.5. Maintenance Testing Of In-Service Equipments

The damaged circuit breaker (CB#17) in Taipower 3rd NPP has been put into service for 20 years at the time of event. Maintenance testing history showed that insulation condition of this circuit breaker was good prior to the event however that being the case the circuit breaker should not have exploded when faced by transient voltage no higher than its BIL of 60kV. This shows that the current diagnostic method of insulation degradation (insulation resistance measurement, dielectric power factor measurement) may not be sensitive enough to detect insulation degradation due to ageing or repeated VFTO strikes. It is recommended that the reliability of such tests, including both the tool used, methodology employed, and interpretation of testing results (including monitoring the trend of measurement results) be further improved. For equipment subject to repeated switching surges, a higher standard should be applied.

5. CONCLUSION

Most NPP's in the world have been designed in such way that their local power loads are provided by "multiple independent sources" to ensure continuous power supply even during faulted periods. However, unless the NPP's local power grid is properly configured and its protection system properly designed, all these "multiple independent sources" can failed at same time as exemplified by Taipower's "318 Event". In view of the many similarities in design and other risk factors for world NPP's, it is of utmost importance that the lessons learned from Taipower's 3rd NPP "318 Event" be properly addressed.

This Chapter examines the Taipower "318 Event" in detail to demonstrate the various possibilities that could leads to NPP blackout. The possibilities investigated include: NPP's location factor, NPP local power grid configuration, cable parameters, switching events, switching surges propagating to MV circuits, ferroresonances, remote tripping, and manual starting difficulties. The lessons learned and proposed countermeasures are summarized in the previous section.

In summary, to ensure the proper design of NPP power protection system, the following 3 considerations must be incorporated:

1 **Check Independence of Equipment and Protection Zone for Various Scenarios:** The "318 Event" was caused by a single CB failure (CB#17) but leading to a complete NPP blackout for over 2 hours. This is mainly due to (1) the breakdown of CB#17 took down CB#15 at the same time due to their physical proximity. (2) The bus configuration cause none of the 2nd, 3rd, or 4th backup power to be available when both CB#15 and #17 both fails and CB#15 created a permanent line to ground fault. (3) The last resort (DG5) was located in a building filled with smoke caused by the CB#17's breakdown making manual starting extremely challenging. (4) The sustained overvoltage in the system could have been avoided should the tripping of the EHV cable be done on both ends of the line instead of just the remote end. All of the above suggest that the independence of equipment and protection zone have failed and needs to be taken into consideration when improving existing or future designs.

2 **Accumulated Equipment Stress Must be Monitored and Considered**: Particularly relevant for NPP's located on the seashore and subject to frequent line switching, equipment stress manifested in the form of reduced insulation level must be subject to more frequent and detailed monitoring. This would include not only absolute value measuring but also trending the measurement so that early signs of equipment weakness can be identified and proper measures be adopted to address it.

3 **Use System Protection Design Approach**: The cause of "motor generating effect", "neutral voltage transfer", "VFTO", and "ferroresonance" occurring during the "318 Event" cannot be addressed one by one and need to be taken into consideration from a system protection perspective. This would include the consideration for using different tripping scheme (such as Direct Transfer Trip on the EHV line), adding additional protection device (such as installing surge absorbers on the MV bus), as well as reconfiguring the bus connections.

ACKNOWLEDGEMENTS

The authors wish to thank Professor He Zhao, Advisor to China Electric Power Research Institute, and Dr. Edward Hsi, EE Department, Chung Yuan Christian University, for their valuable comments on this work.

REFERENCES

1. C. H. Lee, S. C. Hsu, P. H. Hsi, S. L. Chen, of. V. F. T. O. Transferring, E. H. V. from, M. V. to, Observed. System, Taiwan's. in, No, Nuclear Power Plant. in IEEE Trans. Power Delivery 2011262

2. W. Jakel, A. B. Muller, Switching Transient Levels Relevant to Medium Voltage Switchgear and Associated Instrumentation. in Proc. International Conference and Exhibition on ElectroMagnetic Compatibility, June 12131999

3. W. Buesch, J. Marmonier, G. Palmieri, O. Chuniaud, M. G. I. S. Miesch, Transformers. E. M. C. Instrument, Tests. Conformity, a. for, Operation. Reliable, an. in, Substation. Upgraded, in an Upgraded Substation. in Proc. Conference on Electric Power Supply Industry, Oct. 23272000

4. Uglesic, S. Hutter, V. Milardic, I. Ivankovic, B. Filiovic-Grcic, Electromagnetic Disturbances of the Secondary Circuits in Gas Insulated Substation due to Disconnector Switching. in Proc. International Conference on Power Systems Transients, Sep. 28Oct. 2, 2003

5. Atomic Energy Council, The Station Blackout Incident of the Maanshan NPP unit 1: 2001.

6. T. P. Tsao, C. C. Ning, Analysis of Ferroresonant Overvoltage at Maanshan Nuclear Power Station in Taiwan. IEEE Transaction on Power Delivery 2006212

7. Greenwood, Transients. Electrical, Power. in, Systems, John Wiley & Sons, Inc.

8. J. C. Das, Surges transferred through transformers. in Proc. 2002Annual Pulp and Paper Industry Technical Conference, 1721June, 2002.: 139-147

9. S. L. A. Chen, on. study, Feasibility. The, Install. to, Absorber. Surge, Low. at, Side. Voltage, 3. of, k. V. , 1. , k. V. Start-up, in. Transformer, Nuclear. The 3rd, Plant. Power, Taiwan Power Company, Research Report, TPC-546-91-2104-10: 2003

10. H. Zhao, personal, communication: 2005

11. CIGRE WG 33/13-09 Very Fast Transient Phenomena Associated with Gas Insulated Substations, CIGRE Report: 1988.

12. J. Meppeline, K. J. Diederich, K. Feser, W. R. Pfaff, Very fast transients in GIS. IEEE Trans. Power Delivery 198941

13. N. Fujimoto, S. A. Boggs, of. G. I. S. Characteristics, Short. Disconnector-Induced, Transients. Risetime, on. Incident, Connected. Externally, System. Power, I. E. E. Components, IEEE Trans. Power Delivery 19883

14. V. V. Kumar, J. M. Thomas, M. S. Naidu, Influence of Switching Conditions on The VFTO Magnitudes in a GIS. IEEE Trans. Poer Delivery 2001164

15. M. Popov, L. Sluis, R. P. P. van der Smeets, J. L. Roldan, Analysis of Very Fast Transients in Layer-Type Transformer Windings. IEEE Trans. Power Delivery 200722

16. S. Fujita, Y. Shibuya, M. Ishii, of. V. F. T. Influence, Shell. on-Type, I. E. E. E. Transformer, Power. Trans, Delivery, 2007221

17. Y. Shibuya, S. Fujita, E. Tamaki, Analysis of Very Fast Transients in Transformers. in Proc. IEE Generation, Transmission and Distribution Conference 20011485

18. S. Ogawa, E. Haginomori, S. Nishiwaki, T. Yoshiida, K. Terasaka, Estimation of Restriking Transient Overvoltage on Disconnecting Switch for GIS. IEEE Trans. Power System 198612

19. M. M. Rao, M. J. Thomas, D. P. Singh, Frequency Characteristics of Very Fast Transient Currents in a 245 kV GIS. IEEE Trans. Power Delivery 2005204

20. R. P. P. Smeets, W. A. Linden, M. van der Achterkamp, G. C. Pamstra, E. M. Meulemeester, De Disconnector Switching in GIS: Three-Phase Testing and Phenomena. IEEE Trans. Power Delivery 2000151

21. J. Christian, j. Xie, Very Fast Transient Oscillation Measurement at Three Gorges Left Bank Hydro Power Plant. in Proc. 2006International Conference on Power System Technology, 2226Oct. 2006.: 1-7.

22. L. Y. Ji, W. H. Huang, Z. Y. Zhang, W. Shi, Analysis and Simulation of Conducted Interference in Three-Phase in One tank GIS. in Proc. 2009Second Asia-Pacific Conference on Computational Intelligence and Industrial Applications.: 269299

23. C. H. Lee, Simulation and Analysis of The Very Fast Transient Overvoltage in Medium Voltage Systems. Ph. D Thesis, National Tsing Hua University, Taiwan: 2011

Chapter 6

Reliability of Passive Systems in Nuclear Power Plants

Luciano Burgazzi

[1] Reactor Safety and Fuel Cycle Methods Technical Unit, ENEA, Italian National Agency for New Technologies, Energy and Sustainable Economic Development, Bologna,, Italy

1. INTRODUCTION

In order to tackle the development of advanced nuclear technologies, the reliability of passive systems has become an important subject and area under discussion, for their extensive use in new and advanced nuclear power plants, (NEA, 2002), in combination with active safety or operational systems.

Following the IAEA definitions, [1], a passive component does not need any external input or energy to operate and it relies only upon natural physical laws (e.g. gravity, natural convection, conduction, etc.) and/or on inherent characteristics (properties of materials, internally stored energy, etc.) and/or 'intelligent' use of the energy that is inherently available in the system (e.g. decay heat, chemical reactions etc.).

The term "passive" identifies a system which is composed entirely of passive components and structures or a system which uses active components in a very limited way to initiate subsequent passive operation. That is why passive systems are expected to combine among others, the advantages of simplicity, a decrease in the need for human interaction and a reduction or avoidance of external electrical power or signals. These attractions may lead to increased safety and acceptability of nuclear power generation if the detractions can be reduced.

Besides the open feedback on economic competitiveness, special aspects like lack of data on some phenomena, missing operating experience over the wide range of conditions, and driving forces which are smaller - in most cases - than in active safety systems, must be taken into account: the less effective performance as compared to active safety systems has a strong impact on the reliability assessment of passive safety systems.

A categorisation has been developed by the IAEA in [1] distinguishing:

- physical barriers and static structures (e.g. pipe wall, concrete building).

This category is characterized by:

- no signal inputs of "intelligence", no external power sources or forces,
- no moving mechanical parts,
- no moving working fluid.

Examples of safety features included in this category are physical barriers against the release of fission products, such as nuclear fuel cladding and pressure boundary systems; hardened building structures for the protection of a plant against seismic and or other external events; core cooling systems relying only on heat radiation and/or conduction from nuclear fuel to outer structural parts, with the reactor in hot shutdown; and static components of safety related passive systems (e.g., tubes, pressurizers, accumulators, surge tanks), as well as structural parts (e.g., supports, shields).

- moving working fluids (e.g. cooling by free convection). This category is characterized by:

- no signal inputs of "intelligence", no external power sources or forces,
- no moving mechanical parts, but
- moving working fluids.

Examples of safety features included in this category are reactor shutdown/emergency cooling systems based on injection of borated water produced by the disturbance of a hydrostatic equilibrium between the pressure boundary and an external water pool; reactor emergency cooling systems based on air or water natural circulation in heat exchangers immersed in water pools (inside containment) to which the decay heat is directly transferred;

containment cooling systems based on natural circulation of air flowing around the containment walls, with intake and exhaust through a stack or in tubes covering the inner walls of silos of underground reactors; and fluidic gates between process systems, such as "surge lines" of Pressurized Water Reactors (PWRs).

- .moving mechanical parts (e.g. check valves).

This category is characterized by:

- no signal inputs of "intelligence", no external power sources or forces; but
- moving mechanical parts, whether or not moving working fluids are also present.

Examples of safety features included in this category are emergency injection systems consisting of accumulators or storage tanks and discharge lines equipped with check valves; overpressure protection and/or emergency cooling devices of pressure boundary systems based on fluid release through relief valves; filtered venting systems of containments activated by rupture disks; and mechanical actuators, such as check valves and spring-loaded relief valves, as well as some trip mechanisms (e.g., temperature, pressure and level actuators).

- external signals and stored energy (passive execution/active actuation, e.g. scram systems).

This category addresses the intermediary zone between active and passive where the execution of the safety function is made through passive methods as described in the previous categories except that internal intelligence is not available to initiate the process. In these cases an external signal is permitted to trigger the passive process. To recognize this departure, this category is referred to as "passive execution/active initiation".

Examples of safety features included in this category are emergency core cooling and injections systems based on gravity that initiate by battery-powered electric or electro-pneumatic valves; emergency reactor shutdown systems based on gravity or static pressure driven control rods.

According to this classification, safety systems are classified into the higher categories of passivity when all their components needed for safety are passive. Systems relying on no external power supply but using a dedicated, internal power source (e.g., a battery) to supply an active component are not subject to normal, externally caused failures and are included in the lowest category of passivity. This kind of system has active and passive characteristics at different times, for example, the active opening of a valve initiates subsequent passive operation by natural convection.

Inclusion of failure modes and reliability estimates of passive components for all systems is recommended in probabilistic safety assessment (PSA)[1] - studies. Consequently the reliability assessment of passive safety systems, defined as the probability to perform the requested mission to achieve the generic safety function, becomes an essential step.

Notwithstanding that passive systems are credited a higher reliability with respect to active ones, – because of the smaller unavailability due to hardware failure and human error -, there is always a nonzero likelihood of the occurrence of physical phenomena leading to pertinent failure modes, once the system comes into operation. In fact the deviations of the natural forces or physical principles, upon which they rely, from the expected conditions can impair the performance of the system itself. This remark is especially applicable to type B passive systems (i.e. implementing moving working fluids) named thermal-hydraulic passive systems, due to the small engaged driving forces and the thermal-hydraulic phenomena affecting the system performance.

Indeed, while in the case of passive A systems the development of the structural reliability analysis methodology can be carried out with the application of the principles of the probabilistic structural mechanics theory, and operating experience data can be inferred for the reliability assessment of passive C and D components, there is yet no agreed approach as far as passive B systems are concerned.

In fact, such passive safety systems in their designs rely on natural forces, such as gravity or natural convection, to perform their accident prevention and mitigation functions once actuated and started: these driving forces are not generated by external power sources (e.g., pumped systems), as is the case in operating reactor designs. Because the magnitude of the natural forces, which drive the operation of passive systems, is relatively small, counter-forces (e.g. friction) can be of comparable magnitude and cannot be ignored as it is generally the case of systems including pumps. Moreover, there are considerable uncertainties associated with factors on which the magnitude of these forces and counter forces depends (e.g. values of heat transfer coefficients and pressure losses). In addition, the magnitude of such natural driving forces depends on specific plant conditions and configurations which could exist at the time a system is called upon to perform its safety function. All these aspects affect the thermal-hydraulic (T-H) performance of the passive system.

Consequently, a lot of efforts have been devoted mostly to the development of consistent approaches and methodologies aimed at the reliability assessment of the T-H passive systems, with reference to the evaluation of the implemented physical principles (gravity, conduction, etc.). For example, the system fault tree in case of passive systems would consist of basic events, representing failure of the physical phenomena and failure of activating devices: the use of thermal-hydraulic analysis related information for modeling the passive systems should be considered in the assessment process.

The efforts conducted so far to deal with the passive safety systems reliability, have raised an amount of open issues to be addressed in a consistent way, in order to endorse the proposed approaches and to add credit to the underlying models and the eventual reliability figures, resulting from their application. In fact the applications of the proposed methodologies are to a large extent dependent upon the assumptions underlying the methods themselves. At the international level, for instance, IAEA recently coordinated a research project, denoted as *"Natural Circulation Phenomena, Modelling and Reliability of Passive Systems"* (2004-2008), [2,3], while another coordinated research project on *"Development of Methodologies for the Assessment of Passive Safety System Performance in Advanced Reactors"* (2008-2011) is currently underway: while focus of the former project has been the natural circulation and related phenomena, the objective of the latter program is to determine a common analysis-and-test method for reliability assessment of passive safety system performance. This chapter provides the insights resulting from the analysis on the technical issues associated with assessing the reliability of passive systems in the context of nuclear safety and probabilistic safety analysis, and a viable path towards the implementation of the research efforts in the related areas is delineated as well. Focus on these issues is very important since it is the major goal of the international research activities (e.g. IAEA) to strive to reach a common consensus about the different proposed approaches. The chapter is organized as follows: after an overview on passive safety systems being implemented in the design of innovative reactors and an introduction on the main components of Probabilistic Safety Assessment approach, at first the current available methodologies are illustrated and compared, the open issues coming out from their analysis are identified and for which one of them the state of the art and the outlook is presented; the relative importance of each of them within the evaluation process is presented as well.

2. PASSIVE SYSTEMS IMPLEMENTATION IN ADVANCED REACTOR DESIGNS

Several advanced water cooled reactor designs incorporate passive safety systems based on natural circulation, as described in [2,3]: some of the most relevant design concepts for natural circulation systems are described hereafter and namely as regards AP600/AP1000, ESBWR and ABWR designs.

It is important to note that the incorporation of systems based on natural circulation to achieve plant safety and economic goals is being extended also to Generation-IV reactor concepts: however due to the early stage of the design - many systems are not yet established - they are not explicitly addressed.

2.1. Ap600/Ap1000 Passive Residual Heat Removal Systems (Prhr)

Figure 1 presents a schematic that describes the connections of the primary system passive safety systems.

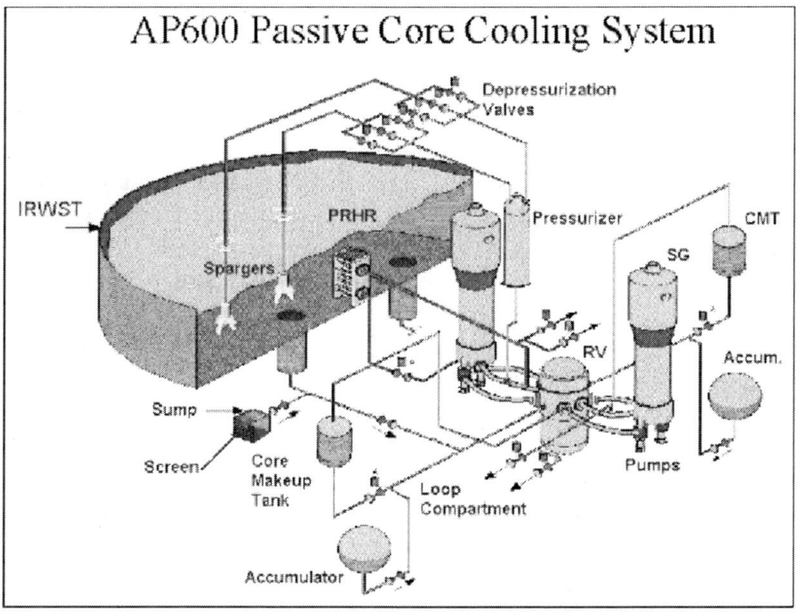

Figure 1. Passive Safety Systems used in the AP600/AP1000 Designs

The AP600/AP1000 passive safety systems consist of:

- A Passive Residual Heat Removal (PRHR) System

- Two Core Make-up Tanks (CMTs)
- A Four Stage Automatic Depressurization System (ADS)
- Two Accumulator Tanks (ACC)
- An In-containment Refueling Water Storage Tank, (IRWST)
- A Lower Containment Sump (CS)
- Passive Containment Cooling System (PCS)

The PRHR implemented in the Westinghouse AP1000 design consists of a C-Tube type heat exchanger in the water-filled In-containment Refuelling Water Storage Tank (IRWST) as shown in the schematic given in Figure 2. The PRHR provides primary coolant heat removal via a natural circulation loop. Hot water rises through the PRHR inlet line attached to one of the hot legs. The hot water enters the tube sheet in the top header of the PRHR heat exchanger at full system pressure and temperature. The IRWST is filled with cold borated water and is open to containment heat removal from the PRHR heat exchanger occurs by boiling on the outside surface of the tubes. The cold primary coolant returns to the primary loop via the PRHR outline line that is connected to the steam generator lower head.

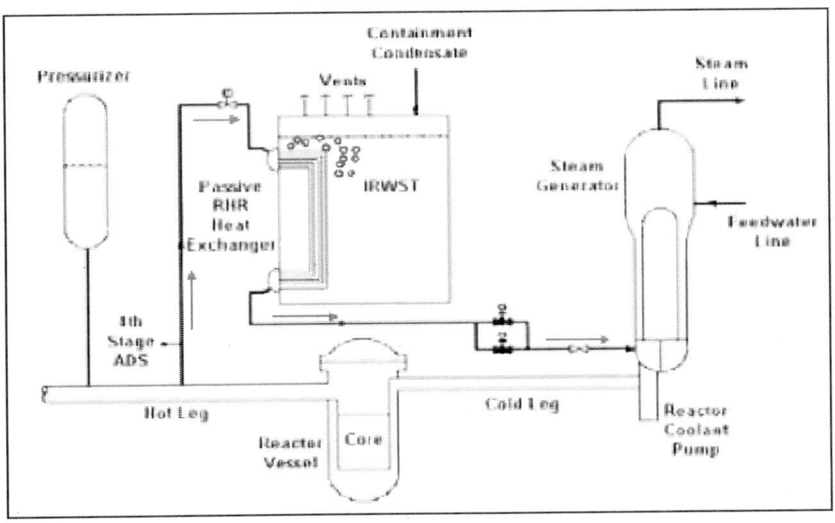

Figure 2. AP1000 passive residual heat removal systems (PRHR)

2.2. Esbwr (Economic Simplified Boiling Water Reactor) Isolation Condenser System (Ics)

During a Loss of Coolant Accident (LOCA), the reactor shuts down and the Reactor Pressure Vessel (RPV) is isolated by closing the main steam line isolation valves. The ICS removes decay heat after any reactor isolation. In other words, the ICS passively removes sensible and core decay heat from the reactor when the normal heat removal system is unavailable. Decay heat removal limits further increases in steam pressure and keeps the RPV pressure below the safety set point. The arrangement of the IC heat exchanger is shown in Figure 3.

The ICS consists of four independent loops, each containing two heat exchanger modules that condense steam inside the tube and transfers heat by heating/evaporating water in the IC pool, which is vented to the atmosphere. This transferring mechanism from IC tubes to the surrounding IC pool water is accomplished by natural convection, and no forced circulation equipment is required.

The ICS is initiated automatically by any of the following signals: high reactor pressure, main steam line isolation valve (MSIV) closure, or an RPV water level signal. To operate the ICS, the IC condensate return valve is opened whereupon the standing condensate drains into the reactor and the steam water interface in the IC tube bundle moves downward below the lower headers.

2.3. Esbwr Passive Containment Cooling System (Pccs)

The PCCS is a passive system which removes the decay heat released to the containment and maintains the containment within its pressure limits for design basis accidents such as a LOCA. The schematic of the PCCS is shown in Figure 4. The PCC heat exchangers receive a steam-gas mixture from the Dry Well (DW), condense the steam and return the condensate to the RPV via the Gravity Driven Cooling System GDCS pools. The non condensable gas is vented to the Wet Well (WW) gas space through a vent line submerged in the Suppression Pool (SP). The venting of the non condensable gas is driven by the differential pressure between the DW and WW. The PCCS condenser, which is open to the containment, receives a steam-gas mixture supply directly from the DW. Therefore, the PCCS operation requires no sensing, control, logic or power actuated devices for operation. The PCCS consists of six PCCS condensers. Each PCCS condenser is made of two identical modules and each entire PCCS condenser two-module assembly is designed for 11 MWt capacity. The condenser condenses steam on the tube side and transfers heat to the water in the IC/PCC pool. The evaporated steam in the IC/PCC pool is vented to the atmosphere. PCCS condensers are located in the large open IC/PCC pool, which are designed to allow full use of the collective water inventory.

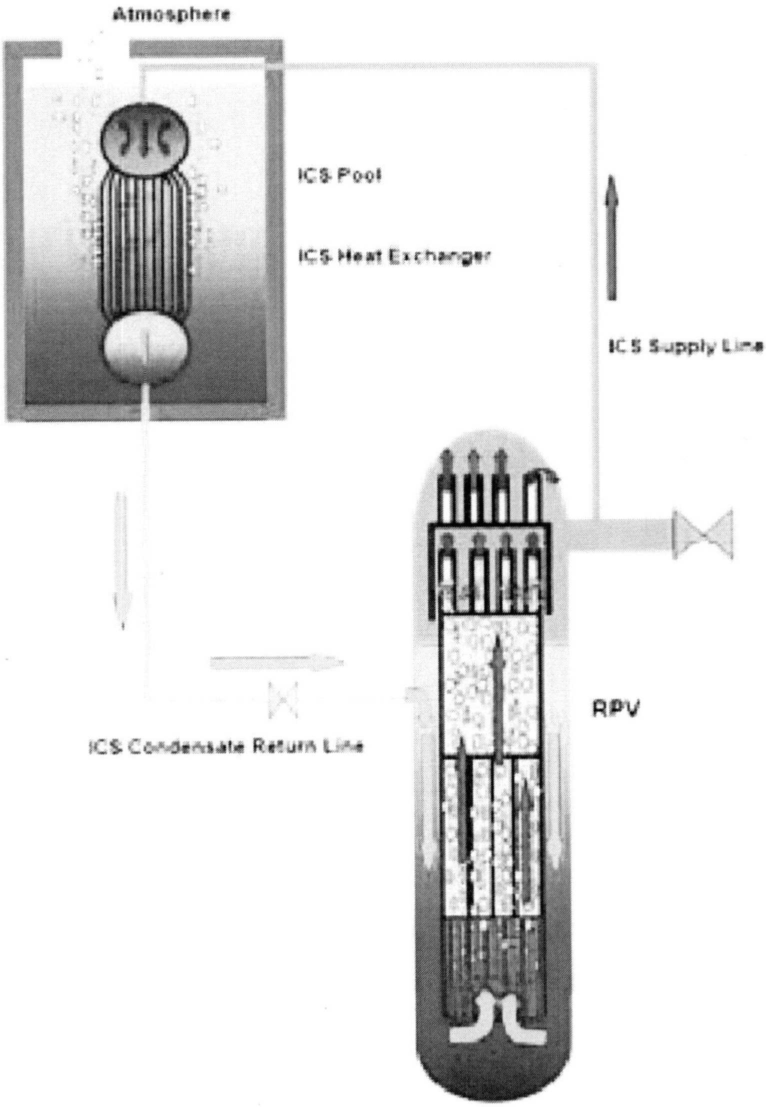

Figure 3. Isolation condenser arrangement

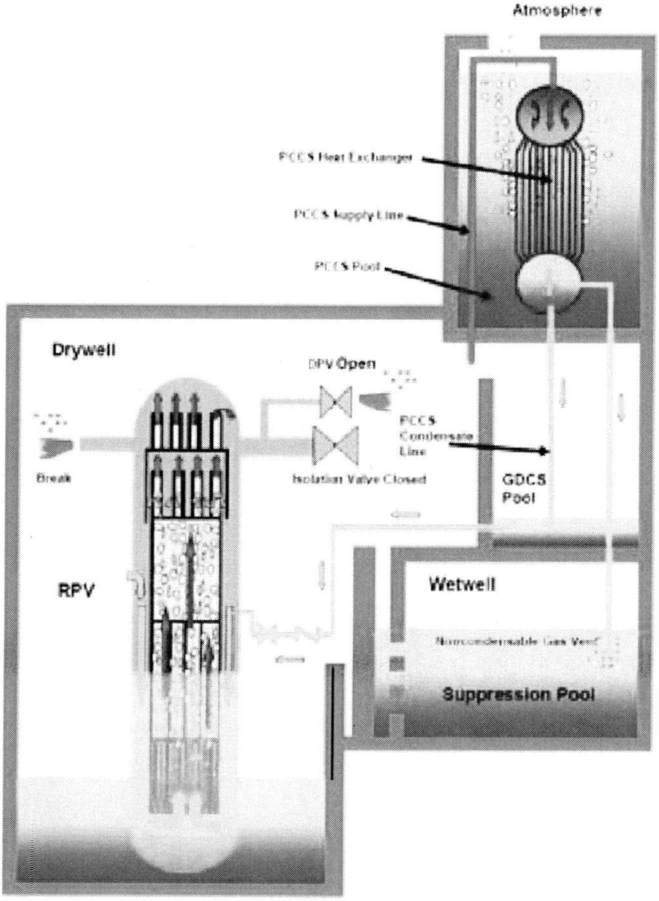

Figure 4. Passive containment cooling condenser arrangement

2.4. Abwr (Advanced Boiling Water Reactor) Passive Reactor Cooling System And Passive Containment Cooling System

The passive heat removal system (PHRS) consists of two dedicated systems (Figure 5, right) namely the passive reactor cooling system (PRCS: the same as Isolation condenser) and the passive containment cooling system (PCCS), that use a common heat sink pool above the containment allowing a one-day grace period, with a 4*50% redundancy (Figure 5, left). These passive systems not only cover beyond DBA condition, but also provide in-depth heat removal backup for the RHR.

In addition, they provide the overpressure protection safety function, practically excluding the necessity of containment venting before and after core damage. Figure 6 shows PCCS functional schematic and an example of containment pressure transient following typical low pressure core melt scenario.

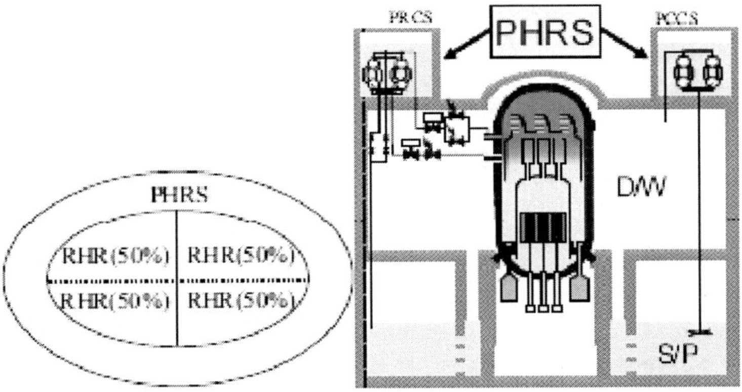

Figure 5. ABWR Passive heat removal system

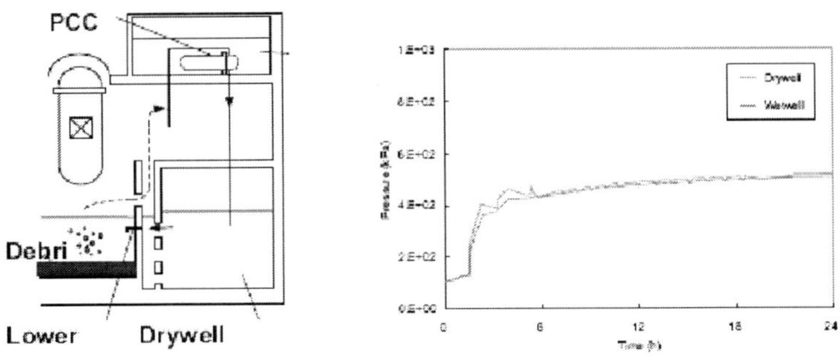

Figure 6. Example of containment pressure transient following typical low pressure core melt scenario.

3. OVERVIEW OF PSA

PSA methodology widely used in the nuclear power industry is deemed helpful to the safety assessment of the facility and along the correspondent licensing process: probabilistic safety assessment can provide insights into safety and identify measures for informing designers of the safety of the plant.

The first comprehensive application of the PSA dates back to 1975, to the United States Nuclear Regulatory Commission's (U.S. NRC) Reactor Safety Study [4]. Since that pioneering study, there has been substantial methodological development, and PSA techniques have become a standard tool in the safety evaluation of the nuclear power plants (NPPs) and industrial installations in general. Due to historical reasons, the PSA sometimes is called PRA.

As the most important area of PSA projects remains nuclear power plants, mainly due to the specific features of the nuclear installations, three levels of PSA have evolved:

Level 1: The assessment of plant failures leading to core damage and the estimation of core damage frequency. A Level 1 PSA provides insights into design weaknesses and ways of preventing core damage. In the case of other industrial assessments, Level 1 PSA provides estimates of the accidents frequency and the main contributors.

Level 2: As possible releases are additionally protected by containment in most NPPs, PSA at this response and severe accident management possibilities. The results obtained in Level 1 are the basis for Level 2 quantification. In the case of other industrial assessments, Level 2 PSA might be fully covered by Level 1, as containment function is rather unique feature and is not common in other industries.

Level 3: The assessment of off-site consequences leading to estimates of risks to the public. Level 3 incorporates results om both previous levels.

Level1 PSA is the most important level and creates the background for further risk assessment, therefore it will be presented in detail. The structure of the other levels is much more application specific, and will be discussed only in general.

The methodology is based on systematically: 1) postulating potential accident scenarios triggered by an initiating event (IE), 2) identifying the systems acting as "defences" against these scenarios, 3) decomposing the systems into components, associating the failure modes and relative probabilities, 4) assessing the frequency of the accident scenarios. Two elements of the PSA methodology typically stand out:

- The event tree (ET) which is used to model the accident scenarios: it represents the main sequences of functional success and failure of safety systems appointed to cope with the initiating events and the consequences of each sequence. These consequences, denoted also as end states, are identified either as a safe end state or an accident end state.

- The fault tree (FT) which documents the systematic, deductive analysis of all the possible causes for the failure of the required function within an accident scenario modelled by the ET. A FT analysis is performed for each of the safety systems, required in response to the IE.

Assigning the safe end state to a sequence means that the scenario has been successfully terminated and undesired consequences have not occurred. In contrast the accident end state means that the sequence has resulted in undesired consequences.

Synthetically, the methodology embraced for the analysis consists of the following major tasks:

- identification of initiating events or initiating event groups of accident sequences: each initiator is defined by a frequency of occurrence;

- systems analysis: identification of functions to be performed in response to each initiating events to successfully prevent plant damage or to mitigate the consequences and identification of the correspondent plant systems that perform these functions (termed front-line systems): for each system the probability of failure is assessed, by fault tree model;

- accident sequences development by constructing event trees for each initiating event or initiating event groups;

- accident sequences analysis to assess the frequencies of all relevant accident sequences;

- identification of dominant sequences on a frequency-consequence base, i.e. the ones presenting the most severe consequences to the personnel, the plant, the public and the environment and definition of the reference accident scenarios to be further analysed through deterministic transient analysis (for instance by t-h code simulation), in order to verify the fulfilment of the safety criteria. Consequences in the case of Level 1 PSA of NPPs are usually defined as degrees of reactor core damage, including 'safe' state and 'severe' accident state.

One of the main issues encountered in probabilistic analysis concerns the availability of pertinent data for the quantification of the risk, which eventually raises a large uncertainty in the results achieved. Usually these data are accessible from consolidated data bases (e.g. IAEA), resulting from the operational experience of the plants.

They pertain, for instance, to component failure rates, component probability on demand, initiating event frequency: for this reason within a PSA study usually an uncertainty analysis, in addition to a sensitivity analysis, is required in order to add credit to the model and to assess if sequences have been correctly evaluated on the probabilistic standpoint.

Event trees are used for the graphical and logical presentation of the accident sequences. An example of an event tree is shown in Figure 7. The logical combinations of success/failure conditions of functions or systems (usually safety systems, also called front-line systems) in the event tree are modelled by the fault tree.

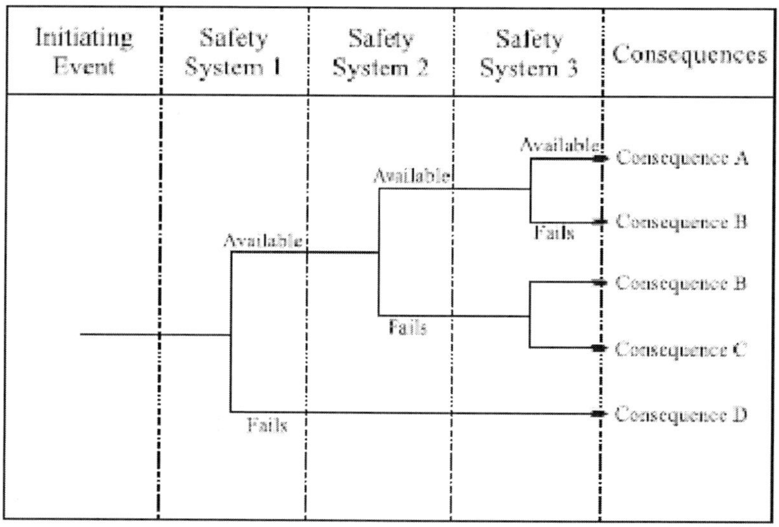

Figure 7. Example of an event tree

A fault tree logically combines the top event (e.g. complete failure of a support system) and the causes for that event (e.g. equipment failure, operator error etc.). An example of the fault tree is shown inFigure 8. The fault tree mainly consists of the basic events (all possible causes of the top event that are consistent with the level of detail of the study) and logical gates (OR, AND, M out of N and other logical operations). Other modelling tools, like common cause failures, house or area events are also used in the fault trees. All front-line and support systems are modelled by the fault trees and then combined in the event trees depending on the initiating event.

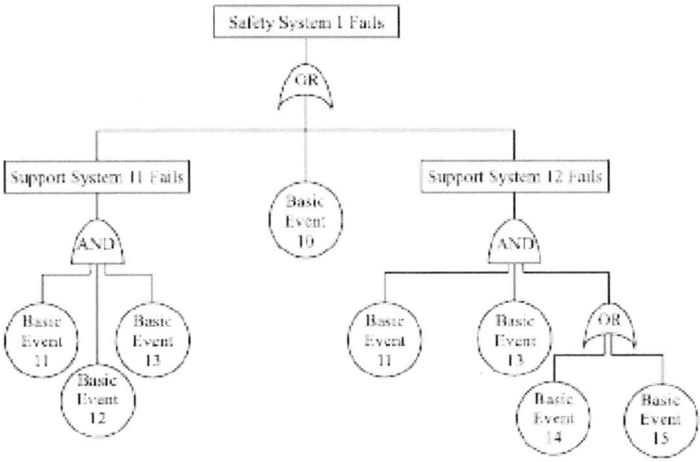

Figure 8. Example of a fault tree

A fault tree is capable to include rather special cases, usually identified in complex systems. These include system and components dependencies, called common cause failures (simultaneous failures of several components due to the same reason), area events (usually fire, flood etc., which damages groups of components in certain rooms), human actions (operator errors or mitigation actions).

The PSA is a powerful tool that can be used in many different ways to assess, understand and manage risk. Its primarily objectives are the following:

- estimate risk level of the facility,
- identify dominant event sequences affecting safety of the facility,
- identify systems, components and human actions important for safety,
- assess important dependencies (among systems or man-machine interactions),
- provide decision support in various application areas.

The growing area of PSA use is extensive support of probabilistic results in risk management and decision-making processes. The main areas of the PSA applications are assessment of design modifications and back-fitting, risk informed optimization of the Technical Specifications, accident management, emergency planning and others. Several modern tools of risk management are also based on the PSA model, such as risk monitoring, precursor analysis and others.

Despite its popularity among the risk assessment tools, the PSA has a number of imitations and drawbacks. The main limitations of the PSA model are the following:

Binary representation of the component state. Only two states are analyzed: failed state or fully functioning state. However, this is not always realistic, as intermediate states are also possible. The same limitation exists for the redundant systems with certain success criteria - system is in failed state (success criteria is not satisfied) or in full power. The intermediate states for redundant systems are even more important.

Independence. In most cases, the components are assumed to be independent (except modelled by CCF), however there are many sources of dependencies, not treated by the model.

Aging effect. The aging effect is ignored because of the constant failure rate assumption. The only conservative possibility to treat the aging impact is to perform sensitivity study.

Time treatment. The FT/ET model is not capable to treat time explicitly during the accident progression. This is one of the major drawbacks of the methodology. In realistic systems, many parameters and functions depend on time and this is not encountered in the model and only approximate chronological order is assumed.

Uncertainty of the calculations. Uncertainties are inevitable in the PSA results and calculations and therefore direct treatment of the quantitative PSA estimates might be misleading. Due to the fact of uncertainties, the qualitative PSA results (identification of dominant accident sequences, comparison of different safety modifications) are of greater importance than quantitative.

4. PASSIVE SYSTEM UNAVAILABILITY MODEL

The reliability of a passive system refers to the ability of the system to carry out a safety function under the prevailing conditions when required and addresses mainly the related performance stability.

In general the reliability of passive systems should be seen from two main aspects:

- systems/components reliability (e.g. piping, valves), as, for instance, the failure to start-up the system operation (e.g. drain valve failure to open)
- physical phenomena reliability, which addresses mainly the natural circulation stability, and the proneness of the system to the failure is dependent on the boundary conditions and the mechanisms needed for maintaining the intrinsic phenomena rather than on component malfunctions.

These two kinds of system malfunction are to to be considered as ET headings, to be assessed by specific FT components, as shown in figures 9 and 10.

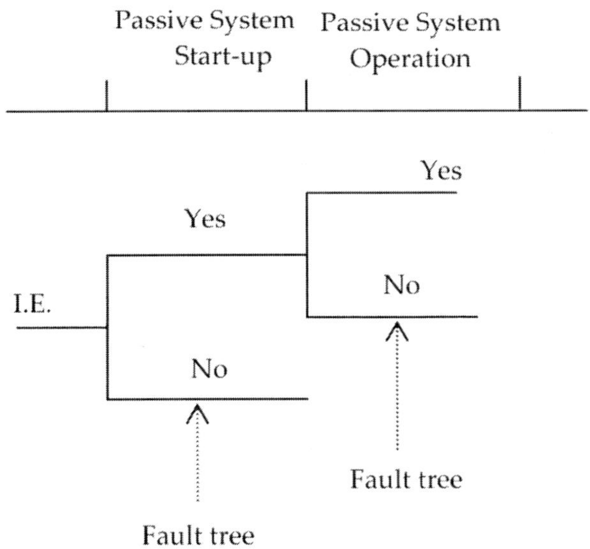

Figure 9. Event tree development

The first facet calls for well-engineered safety components with at least the same level of reliability of the active ones.

The second aspect is concerned with the way the physical principle (gravity and density difference) operate and depends on the surrounding conditions related to accident development in terms of thermal hydraulic parameters evolution (i.e. characteristic parameters as flow rate and exchanged heat flux). This could require not a unique unreliability figure, but the unreliability to be re evaluated for each sequence following an accident initiator, or at least for a small group of bounding accident sequences, enveloping the ones chosen upon similarity of accident progress and expected consequences: with this respect thermal hydraulic analysis of the accident is helpful to estimate the evolution of the parameters during the accident progress.

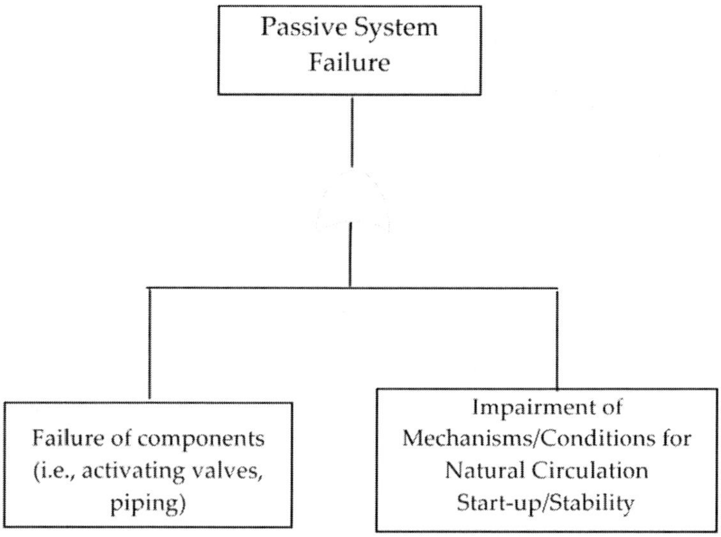

Figure 10. Fault tree model

First step of the analysis is the identification of the failure modes affecting the natural circulation: for this scope two well structured commonly used qualitative hazard analysis, as Failure Mode and Effect Analysis (FMEA) and HAZard and OPerability analysis (HAZOP), specifically tailored on the topic, by considering the phenomenology typical of natural circulation, are adopted.

This analysis concerns both mechanical components (e.g. valve, piping, heat exchanger) of the system and the natural circulation itself, as "virtual" component and the system under investigation is the aforementioned Isolation Condenser.

FMEA is a bottom-up procedure conducted at component level by which each failure mode in a system is investigated in terms of failure causes, preventive actions on causes, consequences on the system, corrective/preventive actions to mitigate the effects on the system, while the HAZOP procedure considers any parameters characteristic of the system (among pressure, temperature, flow rate, heat exchanged through the HX, opening of the drain valve) and by applying a set of "guide" words, which imply a deviation from the nominal conditions as for instance undesired decrease or increase, determines the consequences of operating conditions outside the design intentions. FMEA and HAZOP analysis are shown in Table 1 and 2 respectively.

The analysis points out several factors leading to disturbances in the Isolation Condenser system; the list of these includes:

- Unexpected mechanical and thermal loads, challenging the primary boundary integrity
- HX plugging
- Mechanical component malfunction, i.e. drain valve
- Non-condensable gas build-up
- Heat exchange process reduction: surface oxidation, thermal stratification, piping layout, etc.

Finally a set of critical parameters direct indicators of the failure of the system is identified; these include:

- Non-condensable fraction
- Undetected leakage
- Valve closure area in the discharge line
- Heat loss
- Piping layout
- HX plugged pipes

Each of these failure mode driving parameters is examined to determine the expected failure probability by defining the range and the probability distribution function pertaining to the parameter. These failure characteristics are then used to develop a probabilistic model to predict the natural circulation failure.

As stated before FT technique seems to be the most suitable mean to quantify the passive system unavailability, once introduced the failure modes in the form of critical parameters elementary basic events, linked following the Boolean algebra rules (AND et OR), or in the form of sub-fault trees. However the introduction of passive safety systems into an accident scenario, in the fashion of a safety or front line system, deserves particular attention. The reason is that its reliability figure depends more on the phenomenological nature of occurrence of the failure modes rather than on the classical component mechanical and electrical faults. This makes the relative assessment process different as regards the system model commonly adopted in the fault tree approach as depicted before.

Table 1. FMEA Table for the Isolation Condenser System

Component	Failure Mode	Causes	Prev. Actions on Causes	Consequences	Corrective/Preventive Action on Consequences	Comment
System piping	Rupture	Material defects and aging; Corrosion; Abnormal operation conditions; Vibrations; Local. Stresses; Impact of heavy loads (missile)	Adequate welding process quality; Water chemistry control; In Service inspect; Design against missile generation	LOCA in the Drywell; Instantaneous loss of natural circulation; Emptying of the circuit; Loss of heat removal capability; Loss of reactor coolant inventory	Isolate the breached loop; Safety relief valves actuation; Automatic reactor depressurisation; Gravity Driven Cooling System actuation;	Includes both steam line and drain line Critical Parameter: Undetected Leakage
	Leak	Material defects and aging; Corrosion; Abnormal operation conditions; Vibrations; Local. stresses	Adequate welding process quality; Water chemistry control; In Service inspect.	Small LOCA in the Drywell; Slow emptying of the circuit and natural circulation arrest for long periods of operation; Reduced heat removal capability	Leak monitoring; Isolate the breached loop; Safety relief valves actuation	Critical Parameter: Undetected Leakage
Tube Bundle of the heat exchangers of the IC	Single pipe rupture	Wearing due to vibration and corrosion	Preventive maintenance; Water chemistry control; Leak monitoring	Release of primary water to the pool; Slow emptying of the circuit and natural circulation arrest for long periods of operation; Reduced heat removal capability	Flow monitoring; Isolate the breached loop; Safety relief valves actuation	Critical Parameter: Undetected Leakage
	Multiple pipe rupture	Wearing due to corrosion, vibration and pressure transient	Preventive maintenance; Water chemistry control; Leak monitoring	Release of primary water to the pool; Natural circulation stop; Emptying of the circuit; Loss of reactor coolant inventory; Loss of heat removal capability	Isolate the breached loop; Safety relief valves actuation; Automatic reactor depressurisation; Gravity Driven Cooling System actuation	Critical Parameter: Undetected Leakage
	Single pipe plugging	Crud in the cooling loop; Foreign object in the cooling loop	Water chemistry control; Yearly test of pipes flow; Preventive maintenance	No consequences		Critical Parameter: HX Plugged Pipes
Tube Bundle of the heat exchangers of the IC	Multiple pipe plugging	Violent pressure and vibration transient detaching large amount of crud from pipes walls.	Water chemistry control; Use of suitable materials for cooling loop pipes; Preventive maintenance	Natural circulation stop; Loss of heat removal capability; Reactor pressure and temperature increase	Safety relief valves actuation	Critical Parameter: HX Plugged Pipes

Component	Failure Mode	Causes	Prev. Actions on Causes	Consequences	Corrective/Preventive Action on Consequences	Comment
Drain valve on the return condensate line	Valve fails to open	Control circuit failure; Loss of electric power to motor; Electric motor failure	Redundancy of control devices; Signal to the operator; In Service inspect.	Non triggering of Isolation Condenser if bypass valve does not operate; Loss of heat removal capability; Reactor pressure and temperature increase	Reactor pressure and temperature control; Safety relief valves actuation; Realignment by the operator; Corrective maintenance	Critical Parameter: Partially Open Valve
	Inadvertent valve closing	Spurious signal; Control circuit failure; Human error	Redundancy of control devices; Signal to the operator; Procedured actions	Natural circulation stop in case bypass valve does not operate; Loss of heat removal capability; Reactor pressure and temperature increase	Reactor pressure and temperature control; Safety relief valves actuation; Realignment by the operator; Corrective maintenance	Critical Parameter: Partially Open Valve
Natural Circulation	Envelope failure	Material defects and aging; Corrosion; Abnormal operation conditions; Vibrations; Local. Stresses; Impact of heavy loads (missile)	Adequate welding process quality; Water chemistry control; In Service inspect; Design against missile generation	LOCA in the Drywell; Instantaneous loss of natural circulation; Emptying of the circuit; Loss of heat removal capability; Loss of reactor coolant inventory	Isolate the breached loop; Safety relief valves actuation; Automatic reactor depressurisation; Gravity Driven Cooling System actuation	Includes both steam line and drain line Critical Parameter: Undetected Leakage
	Cracking	Material defects and aging; Corrosion; Abnormal operation conditions; Vibrations; Local. stresses	Adequate welding process quality; Water chemistry control; In Service inspect.	Small LOCA in the Drywell; Slow emptying of the circuit and natural circulation arrest for long periods of operation; Reduced heat removal capability	Leak monitoring; Isolate the breached loop; Safety relief valves actuation	Critical Parameter: Undetected Leakage
	Modification of surface characteristics	Oxidation; Aerosol deposits	Water chemistry control	Reduction in heat exchange efficiency; Reduced heat removal capability	Flow monitoring	Critical Parameter: Oxide Layer
Natural Circulation	Thermal stratification	Temperature dishomogeneity; Density variations; Onset of local thermal hydraulic phenomena	Process control (pressure, flow, temperature)	Reduction of heat convection; Natural circulation blockage; Loss of heat removal capability; Reactor pressure and temperature increase	Flow monitoring; Reactor pressure and temperature control; Safety relief valves actuation	Critical Parameter: Piping Layout, Heat Loss
	Non condensable build-up	Onset of chemical phenomena; Radiolysis products; Impurities	Water chemistry control (PH, O2, H2)	Reduction in heat exchange efficiency; Reduction of heat convection; Natural circulation blockage; Loss of heat removal capability; Reactor pressure and temperature increase	Flow monitoring; Reactor pressure and temperature control; Purging through vent lines; Safety relief valves actuation	Critical Parameter: Non-Condensable Fraction
	Heat dissipation	Thermal insulation degradation; Inaccurate material assembly	In Service inspect.	Reduction of heat convection; Natural circulation impairment	Flow monitoring;	Critical Parameter: Heat Loss

Table 2. HAZOP Table for the Isolation Condenser System

Guide Word	Deviation	Possible Causes	Consequences	Safeguards/Interlocks	Actions Required
PARAMETER: Flow rate					
More of[1]	High Flow	N/A			
Less of	Low Flow	Modifications of surface characteristics (crud deposition, oxidation); Non-condensable build-up; Thermal stratification; Pipe partial plugging; Pipe leak; HX single pipe plugging; HX single pipe rupture Drain valve partial opening	Natural circulation degradation and reduced heat transfer capability; T and P increase	Safety relief valve actuation; Vent line valve actuation	Corrective maintenance; Operator action
No/None	No Flow	Non-condensable build-up; Thermal stratification; Pipe plugging; Pipe rupture; HX Multiple pipe plugging; HX Multiple pipe rupture; Drain valve closed	Natural circulation stop and loss of heat transfer capability; T and P increase	Safety relief valve actuation; Vent line valve actuation; Automatic Depressurisation System actuation; Gravity Driven Cooling System actuation	Corrective maintenance; Operator action
PARAMETER: Pressure					
More of	High Pressure	Non-condensable build-up; Surface modifications (crud, oxidation); HX tube plugging; HX tube rupture Partial valve opening	Natural circulation degradation and reduced heat transfer capability; T increase	Safety relief valve actuation; Vent line valve actuation	Corrective maintenance; Operator action
Less of[1]	Low Pressure	N/A			
No/None	No Pressure	N/A			
PARAMETER: Drain valve opening					
More of	N/A				
Less of	Reduced Opening	Partial blockage	Natural circulation degradation and reduced heat transfer capability; T and P increase	Safety relief valve actuation;	Corrective maintenance; Operator action
No/None	No Opening	Loss of electrical power; Circuit control failure; Electrical motor failure; Valve stuck	Natural circulation stop and loss of heat transfer capability; T and P increase	Safety relief valve actuation; Vent line valve actuation; Automatic Depressurisation System actuation; Gravity Driven Cooling System actuation	Corrective maintenance; Operator action
PARAMETER: Exchanged heat flux					
More of[1]	High Flux	N/A			
Less of	Low Flux	Non-condensable build-up; Surface modifications (crud, oxidation); HX single tube plugging; HX single tube rupture	Natural circulation degradation and reduced heat transfer capability; T and P increase	Safety relief valve actuation; Vent line valve actuation	Corrective maintenance; Operator action
No/None	No Flux	Non-condensable build-up; HX multiple tube plugging; HX multiple tube rupture	Natural circulation stop and loss of heat transfer capability; T and P increase	Safety relief valve actuation; Vent line valve actuation; Automatic Depressurisation System actuation; Gravity Driven Cooling System actuation	Corrective maintenance; Operator action

[1] This deviation is not evaluated, even if it implies an overcooling of the system that could potentially induce to thermal stresses on core structures and reactor components, like the heat exchanger.

In fact, since the failure of the physical process is addressed, the conventional failure model associated with the basic events (i.e. exponential, $e^{-\lambda t}$, λ failure rate, t mission time), commonly used for component failure model, is not applicable: each pertinent basic event will be characterized by defined parameters driving the failure mechanisms - e.g. non-condensable fraction, leak rate, partial opening of the isolation valve, heat exchanger plugged pipes, etc. - and the associated failure criterion. Thus each basic event model pertaining to the relevant failure mode requires the assignment of both the probability distribution and range of the correspondent parameter and the definition of the critical interval defining the failure (for example failure for non-condensable fraction >x%, leak rate > x gr./sec or crack size > x cm^2 and so on).

In order to evaluate the overall probability of failure of the system, the single failure probabilities are combined according to:

$$Pe_t = 1.0- ((1.0 - Pe_1)*(1.0 - Pe_2)*...*(1.0 - Pe_n)) \quad (1)$$

where:

Pe_t overall probability of failure

Pe_1 through Pe_n individual probabilities of failure pertaining to each failure mode, assuming mutually non-exclusive independent events

The failure model relative to each single basic event is given by:

$$Pe_i= \int p_i(x)\, dx \quad x>x_o \quad (2)$$

$p_i(x)$probability distribution function of the parameter x

$x_o$threshold value according to the failure criterion

It's worth noting that the assumed failure criterion, based on the failure threshold for each path, implies the neglecting of the "intermediate" modes of operation of the system or equivalently the degraded performance of the system (up to the failure point): this gives credit for a passive system that "partially works" and has failed for its intended function but provides some operation. This operation could be sufficient to prolong the window for opportunity to recover a failed system, for instance through redundancy configuration, and ultimately prevent or arrest core degradation.

Once the probabilistic distributions of the parameters are assigned, the reliability of the system can be directly obtained from (1) once a failure criterion is assigned and the single failure probabilities are evaluated through

(2): this point is being satisfied by assigning both the range and the probability distributions, basing on expert judgment and engineering assessment. In fact, as further illustrated, difficulties arise in assigning both the range and the probability density functions relative to the critical parameters defining the failure modes, in addition to the definition of a proper failure criterion, because of the lack of operational experience and data.

5. METHODOLOGIES CHARACTERIZATION AND COMPARATIVE ASSESSMENT

A very good description of the various methodologies proposed so far and currently available in the open literature is given in [5].

The earliest significant effort to quantify the reliability of such systems is represented by a methodology known as REPAS (Reliability Evaluation of Passive Systems), [6], which has been developed in late 1990s, cooperatively by ENEA, the University of Pisa, the Polytechnic of Milan and the University of Rome, that was later incorporated in the EU (European Union) RMPS (Reliability Methods for Passive Systems) project. This methodology is based on the evaluation of a failure probability of a system to carry out the desired function from the epistemic uncertainties of those physical and geometric parameters which can cause a failure of the system.

The RMPS methodology, described in [7], was developed to address the following problems: 1) Identification and quantification of the sources of uncertainties and determination of the important variables, 2) Propagation of the uncertainties through thermal-hydraulic (T-H) models and assessment of passive system unreliability and 3) Introduction of passive system unreliability in accident sequence analyses. In this approach, the passive system is modelled by a qualified T-H code (e.g. CATHARE, RELAP) and the reliability evaluation is based on results of code runs, whose inputs are sampled by Monte-Carlo (M-C) simulation. This approach provides realistic assessment of the passive system reliability, thanks to the flexibility of the M-C simulation, which adapts to T-H model complexity without resort to simplifying approximation. In order to limit the number of T-H code runs required by M-C simulation, alternative methods have been proposed such as variance reduction techniques, first and second order reliability methods and response surface methods. The RMPS methodology has been successfully applied to passive systems utilizing natural circulation in different types of reactors (BWR, PWR, and VVER). A complete example of application concerning the passive residual heat removal system of a CAREM reactor is presented in [8]. The RMPS methodology tackles also an important problem, which is the integration of passive system reliability in a

PSA study. So far, in existing innovative nuclear reactor projects PSA's, only passive system components failure probabilities are taken into account, disregarding the physical phenomena on which the system is based, such as the natural circulation. The first attempts performed within the framework of RMPS have taken into account the failures of the components of the passive system as well as the impairment of the physical process involved like basic events in static event tree as exposed in [7]. Two other steps have been identified after the development of the RMPS methodology where an improvement was desirable: the inclusion of a formal expert judgment (EJ) protocol to estimate distributions for parameters whose values are either sparse on not available, and the use of efficient sensitivity analysis techniques to estimate the impact of changes in the input parameter distributions on the reliability estimates.

R&D in the United States on the reliability of passive safety systems has not been as active at least until mid 2000. A few published papers from the Massachusetts Institute of Technology (MIT) have demonstrated their development of approaches to the issue. Their technique has examined TH uncertainties in passive cooling systems for Generation IV-type gas-cooled reactors. The MIT research on the reliability of passive safety systems has taken a similar approach but has focused on a different set of reactor technologies. Their research has examined thermal hydraulic uncertainties in passive cooling systems for Generation IV gas-cooled reactors, as described in [9,10]. Instead of post-design probabilistic risk analysis for regulatory purposes, the MIT research seeks to leverage the capabilities of probabilistic risk assessment (PRA) to improve the design of the reactor systems early in their development life cycle.

In addition to the RMPS approach, a number of alternative methodologies have been investigated for the reliability assessment of T-H passive systems.

Three different methodologies have been proposed by ENEA (Italian National Agency for New Technologies, Energy and Sustainable Economic Development). In the first methodology [11], the failure probability is evaluated as the probability of occurrence of different independent failure modes, a priori identified as leading to the violation of the boundary conditions or physical mechanisms needed for successful passive system operation.

This approach based on independent failure modes introduces a high level of conservatism as it appears that the probability of failure of the system is relevantly high, because of the combination of various modes of failure as in a series system, where a single fault is sufficient to challenge the system performance. The correspondent value of probability of failure can be conservatively assumed as the upper bound for the unavailability of the system, within a sort of "parts-count" reliability estimation.

In the second, [12], modelling of the passive system is simplified by linking to the modelling of the unreliability of the hardware components of the system: this is achieved by identifying the hardware failures that degrade the natural mechanisms upon which the passive system relies and associating the unreliability of the components designed to assure the best conditions for passive function performance.

Thus, the probabilities of degraded physical mechanisms are reduced to unreliability figures of the components whose failures challenge the successful passive system operation. If, on the one hand, this approach may in theory represent a viable way to address the matter, on the other hand, some critical issues arise with respect to the effectiveness and completeness of the performance assessment over the entire range of possible failure modes that the system may potentially undergo and their association to corresponding hardware failures. In this simplified methodology, degradation of the natural circulation process is always related to failures of active and passive components, not acknowledging, for instance, any possibility of failure just because of unfavourable initial or boundary conditions. In addition, the fault tree model adopted to represent the physical process decomposition is used as a surrogate model to replace the complex T-H code that models the system behaviour. This decomposition is not appropriate to predict interactions among physical phenomena and makes it extremely difficult to realistically assess the impact of parametric uncertainty on the performance of the system.

The third approach is based on the concept of functional failure, within the reliability physics framework of load-capacity exceedance [7,13,14]. The functional reliability concept is defined as the probability of the passive system failing to achieve its safety function as specified in terms of a given safety variable crossing a fixed safety threshold, leading the load imposed on the system to overcome its capacity. In this framework, probability distributions are assigned to both safety functional requirement on a safety physical parameter (for example, a minimum threshold value of water mass flow required to be circulating through the system for its successful performance) and system state (i.e., the actual value of water mass flow circulating), to reflect the uncertainties in both the safety thresholds for failure and the actual conditions of the system state. Thus the mission of the passive system defines which parameter values are considered a failure by comparing the corresponding pdfs according to defined safety criteria. The main drawback in the last method devised by ENEA lies in the selection and definition of the probability distributions that describe the characteristic parameters, based mainly on subjective/engineering judgment.

Every one of three methods devised by ENEA shares with the main RMPS approach the issue related to the uncertainties affecting the system performance assessment process. With respect to the RMPS a greater simplicity is introduced, although detrimental to the relevance of the approaches themselves: this is particularly relevant as far as the approach based on hardware components failure is concerned.

Finally a different approach is followed in the APSRA (Assessment of Passive System ReliAbility) methodology developed by BARC (Bhabha Atomic Research Centre, India), see [15]. In this approach, a failure surface is generated by considering the deviation of all those critical parameters, which influence the system performance. Then, the causes of deviation of these parameters are found through root diagnosis. It is attributed that the deviation of such physical parameters occurs only due to a failure of mechanical components such as valves, control systems, etc. Then, the probability of failure of a system is evaluated from the failure probability of these mechanical components through classical PSA treatment. Moreover, to reduce the uncertainty in code predictions, BARC foresee to use in-house experimental data from integral facilities as well as separate.

With reference to the two most relevant methodologies (i.e. RMPS and APSRA), the RMPS consists mainly in the identification and quantification of parameter uncertainties in the form of probability distributions, to be propagated directly into a T-H code or indirectly in using a response surface; the APSRA methodology strives to assess not the uncertainty of parameters but the causes of deviation from nominal conditions, which can be in the failure of active or passive components or systems.

As a result, different approaches are used in the RMPS and APSRA methodologies. RMPS proposes to take into account, in the PSA model, the failure of a physical process. This problem is treated in using a best estimate T-H code plus uncertainty approach. APSRA includes in the PSA model the failure of those components which cause a deviation of the key parameters resulting in a system failure, but does not take into account possible uncertainties on these key parameters. As the consequence, the T-H code is used in RMPS to propagate the uncertainties and in APSRA to build a failure surface. APSRA incorporates an important effort on qualification of the model and use of the available experimental data. These aspects have not been studied in the RMPS, given the context of the RMPS project.

The following Table attempts to identify the main characteristics of the methodologies proposed so far, with respect to some aspects, such as the development of deterministic and probabilistic approaches, the use of deterministic models to evaluate the system performance, the identification of the sources of uncertainties and the application of expert judgment.

Table 3. Main features of the various approaches

Methodology	Probabilistic vs. deterministic	Deterministic Analysis	Uncertainties	Expert Judgment/Experimental data
REPAS/RMPS	Merge of probabilistic and thermal hydraulic aspects	T-H code adopted for uncertainty propagation	Uncertainties in parameters modelled by probability density functions	EJ adopted to a large extent; Statistical analysis when experimental data exist
APSRA	Merge of probabilistic and thermal hydraulic aspects	T-H code adopted to build the failure surface	parameters' deviations from nominal conditions caused by failure of active or passive components (root diagnosis)	Experimental data usage; EJ for root diagnosis
ENEA approaches	Only probabilistic aspects		Uncertainties in parameters	EJ adopted to a large extent (except the approach based on hardware failure)

6. OPEN ISSUES

From the exam of the various methodologies, which have been developed over these most recent years within the community of the safety research, and are currently available in the open literature, the following open questions are highlighted and consequently needs for research in all related areas are pointed out :

- The aspects relative to the assessment of the uncertainties related to passive system performance: they regard both the best estimate T-H codes used for their evaluation and system reliability assessment itself;

- The dependencies among the parameters, mostly T-H parameters, playing a key role in the whole process assessment.

- The integration of the passive systems within an accident sequence in combination with active systems and human actions.

- The consideration for the physical process and involved physical quantities dependence upon time, implying, for instance, the development of dynamic event tree to incorporate the interactions between the physical parameter evolution and the state of the system and/or the transition of the system from one state to another.

It's worth noticing that these two last aspects are correlated, but hey will be treated separately.

- The comparison between active and passive systems, mainly on a functional viewpoint.

All of these points are elaborated in the following, in an attempt to cover the entire spectrum of issues related to the topic, and capture all the relevant aspects to concentrate on and devote resources towards for fulfilling a significant advance.

6.1. Uncertainties

The quantity of uncertainties affecting the operation of the T-H passive systems affects considerably the relative process devoted to reliability evaluation, within a probabilistic safety analysis framework, as recognized in [7].

These uncertainties stem mainly from the deviations of the natural forces or physical principles, upon which they rely (e.g., gravity and density difference), from the expected conditions due to the inception of T-H factors impairing the system performance or to changes of the initial and boundary conditions, so that the passive system may fail to meet the required function. Indeed a lot of uncertainties arise, when addressing these phenomena, most of them being almost unknown due mainly to the scarcity of operational and experimental data and, consequently, difficulties arise in performing meaningful reliability analysis and deriving credible reliability figures. This is usually designated as phenomenological uncertainty, which becomes particularly relevant when innovative or untested technologies are applied, eventually contributing significantly to the overall uncertainty related to the reliability assessment.

Actually there are two facets to this uncertainty, i.e., "aleatory" and "epistemic" that, because of their natures, must be treated differently. The aleatory uncertainty is that addressed when the phenomena or events being modelled are characterized as occurring in a "random" or "stochastic" manner and probabilistic models are adopted to describe their occurrences. The epistemic uncertainty is that associated with the analyst's confidence in the prediction of the PSA model itself, and it reflects the analyst's assessment of how well the PSA model represents the actual system to be modelled. This has also been referred to as state-of-knowledge uncertainty, which is suitable to reduction as opposed to the aleatory which is, by its nature, irreducible. The uncertainties concerned with the reliability of passive system are both stochastic, because of the randomness of phenomena occurrence, and of epistemic nature, i.e. related to the state of knowledge about the phenomena, because of the lack of significant operational and experimental data.

For instance, as initial step, the approach described in [16]. allows identifying the uncertainties pertaining to passive system operation in terms of critical parameters driving the modes of failure, as, for instance, the presence of non-condensable gas, thermal stratification and so on. In this context the critical parameters are recognized as epistemic uncertainties.

The same reference points out, as well, the difference between the uncertainties related to passive system reliability and the uncertainties related to the T-H codes (e.g. RELAP), utilized to evaluate the performance itself, as the ones related to the coefficients, correlations, nodalization, etc.: these specific uncertainties, of epistemic nature, in turn affect the overall uncertainty in T-H passive system performance and impinge on the final sought reliability figure.

A further step of the matter can be found in[11], which attempts to assign sound distributions to the critical parameters, to further develop a probabilistic model. As is of common use when the availability of data is limited, subjective probability distributions are elicited from expert/engineering judgment procedure, to characterize the critical parameters.

Three following classes of uncertainties to be addressed are identified:

- Geometrical properties: this category of uncertainty is generally concerned with the variations between the as-built system layout and the design utilized in the analysis: this is very relevant for the piping layout (e.g. suction pipe inclination at the inlet of the heat exchanger, in the isolation condenser reference configuration) and heat loss modes of failure.

- Material properties: material properties are very important in estimating the failure modes concerning for instance the undetected leakages and the heat loss.

- Design parameters, corresponding to the initial/boundary conditions (for instance, the actual values taken by design parameters, like the pressure in the reactor pressure vessel).

- Phenomenological analysis: the natural circulation failure assessment is very sensitive to uncertainties in parameters and models used in the thermal hydraulic analysis of the system. Some of the sources of uncertainties include but are not limited to: the definition of failure of the system used in the analysis, the simplified model used in the analysis, the analysis method and the analysis focus on failure locations and modes and finally the selection of the parameters affecting the system performance.

The first, second and third groups are part of the category of aleatory uncertainties because they represent the stochastic variability of the analysis inputs and they are not reducible.

The fourth category is referred to the epistemic uncertainties, due to the lack of knowledge about the observed phenomenon and thus suitable for reduction by gathering a relevant amount of information and data. This class of uncertainties must be subjectively evaluated, since no complete investigation of these uncertainties is available.

A clear prospect of the uncertainties as shown in Table 4 [5].

As emphasized above, clearly the epistemic uncertainties address mostly the phenomena underlying the passive operation and the parameters and models used in the T-H analysis of the system (including the ones related to the best estimate code) and the system failure analysis itself. Some of the sources of uncertainties include but are not limited to the definition of failure of the system used in the analysis, the simplified model used in the analysis, the analysis method and the analysis focus of failure locations and modes and finally the selection of the parameters affecting the system performance. With this respect, it is important to underline, again, that the lack of relevant reliability and operational data imposes the reliance on the underlying expert judgment for an adequate treatment of the uncertainties, thus making the results conditional upon the expert judgment elicitation process. This can range from the simple engineering/subjective assessment to a well structured procedure based on expert judgment elicitation, as reported in [17], which outlines the main aspects of the REPAS procedure.

In ref. [17], in order to simplify both the identification of the ranges and their corresponding probabilities, initially discrete values have been selected. As a general rule, a central pivot has been identified, and then the range has been extended to higher and lower values, if applicable. The pivot value represents the nominal condition for the parameter. The limits have been chosen in order to exclude unrealistic values or those values representing a limit zone for the operation demand of the passive system. Once the discrete ranges have been set up, discrete probability distributions have been associated, to represent the probabilities of occurrence of the values. As in the previous step, the general rule adopted is that the higher probability of occurrence corresponds to the nominal value for the parameter. Then lower probabilities have been assigned to the other values, as much low the probability as much wide the distance from the nominal value, as in a sort of Gaussian distribution.

Ultimately, as underlined in the previous section, the methodologies proposed in RMPS and within the studies conducted by MIT address the question by propagating the parameter and model uncertainties, by performing Monte

Carlo simulations on the detailed T-H model based on a mechanistic code, and calculating the distribution of the safety variable and thus the probability of observing a value above the defined limit, according to the safety criterion.

Table 4. Categories of uncertainties associated with T-H passive systems reliability assessment

Aleatory
Geometrical properties
Material properties
Initial/boundary conditions (design parameters)
Epistemic
T-H analysis
Model (correlations)
Parameters
System failure analysis
Failure criteria
Failure modes (critical parameters)

6.2. Dependencies

Alike some other types of analyses for nuclear power plants, the documented experience with PSS reliability seems to focus on the analysis of one passive attribute at a time. In many cases, this may be sufficient, but for some advanced designs with multiple passive features, modelling of the synergistic effects among them is important. For example, modelling of a passive core cooling system may require simultaneous modelling of the amount of non condensable gases which build up along the circuit during extended periods of operation, the potential for stratification in the cooling pool, and interactions between the passive core cooling system and the core. Analysis of each of these aspects independently may not fully capture the important boundary conditions of each system. For instance, with regard to the aforementioned methodologies, the basic simplifying assumption of independence among system performance relevant parameters, as the degradation measures, means that the correlation among the critical parameter distributions is zero or is very low to be judged significant, so that the assessment of the failure probability is quite straightforward. If parameters have contributors to their uncertainty in common, the respective states of knowledge are dependent. As a consequence of this dependence, parameter values cannot be combined freely and independently. Instances of such limitations need to be identified and the dependencies need to be quantified. If the analyst knows of dependencies between parameters explicitly, multivariate distributions or conditional subjective pdfs (probability density functions) may be used. The dependence between the parameters can be also introduced by covariance matrices or by functional relations between the parameters.

As observed in [15], both REPAS and RMPS approaches adopt a probability density function (pdf) to treat variations of the critical parameters considered in the predictions of codes. To apply the methodology, one needs to have the pdf values of these parameters. However, it is difficult to assign accurate pdf treatment of these parameters, which ultimately define the functional failure, due to the scarcity of available data, both on an experimental and operational ground. Moreover, these parameters are not really independent ones to have deviation of their own. Rather deviations of them from their nominal conditions occur due to failure/malfunctioning of other components or as a result of the combination with different concomitant mechanisms. Thus the hypothesis of independence among the failure driving parameters appears non proper.

With reference to the functional reliability approach set forth in [13], the selected representative parameters defining the system performance, for instance coolant flow or exchanged thermal power, are properly modelled through the construction of joint probability functions in order to assess the correspondent functional reliability. A recent study shows how the assumption of independence between the marginal distributions to construct the joint probability distributions to evaluate system reliability adds conservatism to the analysis, [18]: for this reason the model is implemented to incorporate the correlations between the parameters, in the form of bivariate normal probability distributions. That study has the merit to highlight the dependence among the parameters underlying the system performance: further studies are underway, with regard, for instance to the approach based on independent failure modes. As described in the previous section 2, this approach begins by identifying critical parameters, properly modelled through probability functions, as input to basic events, corresponding to the failure modes, arranged in a series system configuration, assuming non-mutually exclusive independent events. It introduces a high level of conservatism as it appears that the probability of failure of the system is relevantly high to be considered acceptable, because of the combination of various modes of failure, where a single fault is sufficient to challenge the system performance. Initial evaluations, [19], reveal that the critical parameters are not suitable to be chosen independently of each other, mainly because of the expected synergism between the different phenomena under investigation, with the potential to jeopardize the system performance. This conclusion allows the implementation of the proposed methodology, by properly capturing the interaction between various failure modes, through modelling system performance under multiple degradation measures. It was verified that when the multiple degradation measures in a system are correlated, an incorrect independence assumption may overestimate the system reliability, according to a recent study, [20].

6.3. Incorporation of Passive System Within Probabilistic Safety Assessment

PSA has been introduced for the evaluation of design and safety in the development of those reactors. A technology-neutral framework, that adopts PSA information as a major evaluation tool, has been proposed as the framework for the evaluation of safety or regulation for those reactors [21,22]. To utilize this framework, the evaluation of the reliability of Passive Systems has been recognized as an essential part of PSA.

In PSA, the status of individual systems such as a passive system is assessed by an accident sequence analysis to identify the integrated behaviour of a nuclear system and to assign its integrated system status, i.e. the end states of accident sequences. Because of the features specific of a passive system, it is difficult to define the status of a passive system in the accident sequence analysis. In other words, the status of a passive system does not become a robust form such as success or failure, since "intermediate" modes of operation of the system or equivalently the degraded performance of the system (up to the failure point) is possible. This gives credit for a passive system that "partially works" and has failed for its intended function but provides some operation: this operation could be sufficient to prolong the window for opportunity to recover a failed system, for instance through redundancy configuration, and ultimately prevent or arrest core degradation [19]. This means that the status of a passive system can be divided into several states, and each status is affected by the integrated behaviour of the reactor, because its individual performance is closely related with the accident evolution and whole plant behaviour.

Ref. [23] lays the foundations to outline a general approach for the integration of a passive system, in the form of a front line system and in combination with active ones and/or human actions, within a PSA framework.

In [7] a consistent approach, based on an event tree representation, has been developed to incorporate in a PSA study the results of reliability analyses of passive systems obtained on specific accident sequences. In this approach, the accident sequences are analyzed by taking into account the success or the failure of the components and of the physical process involved in the passive systems. This methodology allows the probabilistic evaluation of the influence of a passive system on a definite accident scenario and could be used to test the advantage of replacing an active system by a passive system in specific situations.

However in order to generalize the methodology, it is important to take into account the dynamic aspects differently than by their alone modelling into the T-H code. Indeed in complex situations where several safety systems are competing and where the human operation cannot be completely eliminated, this modelling should prove to be impossible or too expensive in computing times. It is thus interesting to explore other solutions already used in the dynamic PSA, like the method of the dynamic event trees, in order to capture the interaction between the process parameters and the system state within the dynamical evolution of the accident.

In the PSA of nuclear power plants (NPPs), accident scenarios, which are dynamic in nature, are usually analyzed with event trees and fault trees.

The current PSA framework has some limitations in handling the actual timing of events, whose variability may influence the successive evolution of the scenarios, and in modelling the interactions between the physical evolution of the process variables (temperatures, pressures, mass flows, etc.,) and the behaviour of the hardware components. Thus, differences in the sequential order of the same success and failure events and the timing of event occurrence along an accident scenario may affect its evolution and outcome; also, the evolution of the process variables (temperatures, pressures, mass flows, etc.,) may affect the event occurrence probabilities and thus the developing scenario. Another limitation lies in the binary representations of system states (i.e., success or failure), disregarding the intermediate states, which conversely concern the passive system operation, as illustrated above.

To overcome the above-mentioned limitations, dynamic methodologies have been investigated which attempt to capture the integrated response of the systems/components during an accident scenario [24].

The most evident difference between dynamic event trees (DETs) and the event trees (ETs) is as follows. ETs, which are typically used in the industrial PSA, are constructed by an analyst, and their branches are based on success/ failure criteria set by the analyst. These criteria are based on simulations of the plant dynamics. On the contrary, DETs are produced by a software that embeds the models that simulates the plant dynamics into stochastic models of components failure. A challenge arising from the dynamic approach to PSA is that the number of scenarios to be analyzed is much larger than that of the classical fault/event tree approaches, so that the a posteriori information retrieval can become quite onerous and complex.

This is even more relevant as far as thermal hydraulic natural circulation passive systems are concerned since their operation is strongly dependent, more than other safety systems, upon time and the state/parameter evolution of the system during the accident progression.

Merging probabilistic models with T-H models, i.e. dynamic reliability, is required to accomplish the evaluation process of T-H passive systems in a consistent manner: this is particularly relevant with regard to the introduction of a passive system in an accident sequence, since the required mission could be longer than 24 h as usual level 1 PSA mission time. In fact for design basis accidents, the passive systems are required to establish and maintain core cooling and containment integrity, with no operator intervention or requirement for a.c. power for 72 h, as a grace time [25].

The goal of dynamic PRA is to account for the interaction of the process dynamics and the stochastic nature/behavior of the system at various stages: it associates the state/parameter evaluation capability of the thermal hydraulic analysis to the dynamic event tree generation capability approach. The methodology should estimate the physical variation of all technical parameters and the frequency of the accident sequences when the dynamic effects are considered. If the component failure probabilities (e.g. valve per-demand probability) are known, then these probabilities can be combined with the probability distributions of estimated parameters in order to predict the probabilistic evolution of each scenario outcome.

A preliminary attempt in addressing the dynamic aspect of the system performance in the frame of passive system reliability is shown in [26], which introduces the T-H passive system as a non-stationary stochastic process, where the natural circulation is modeled in terms of time-variant performance parameters, (as for instance mass flow-rate and thermal power, to cite any) assumed as stochastic variables. In that work, the statistics associated with the stochastic variables change in time (in terms of associated mean values and standard deviations increase or decrease, for instance), so that the random variables have different values in every realization, and hence every realization is different.

6.4. Comparative Assessment Between Active and Passive Systems

The design and development of future water-cooled reactors address the use of passive safety systems, i.e. those characterized by no or very limited reliance on external input (forces, power or signal, or human action) and whose operation takes advantage of natural forces, such as free convection and gravity, to fulfil the required safety function and to provide confidence in the plant's ability to handle transients and accidents. Therefore, they are required to accomplish their mission with a sufficient reliability margin that makes them attractive as an important means of achieving both simplification and cost reduction for future plants while assuring safety requirements with lesser dependence of the safety function on active components like pumps and diesel generators.

On the other hand, since the magnitude of the natural forces, which drive the operation of passive systems, is relatively small, counter-forces (e.g. friction) can be of comparable magnitude and cannot be ignored as is generally the case with pumped systems. This concern leads to the consideration that, despite the fact that passive systems "should be" or, at least, are considered, more reliable than active ones - because of the smaller unavailability due to hardware failure and human error - there is always a nonzero likelihood of the occurrence of physical phenomena leading to pertinent failure modes, once the system enters into operation.

These characteristics of a high level of uncertainty and low driving forces for heat removal purposes justify the comparative evaluation between passive and active options, with respect to the accomplishment of a defined safety function (e.g. decay heat removal) and the generally accepted viewpoint that passive system design is more reliable and more economical than active system design has to be discussed [27].

Here are some of the benefits and disadvantages of the passive systems that should be evaluated vs. the correspondent active system.

- Advantages
- No external power supply: no loss of power accident has to be considered.
- No human factor, implying no inclusion of the operator error in the analysis.
- Better impact on public acceptance, due to the presence of "natural forces".
- Less complex system than active and therefore economic competitiveness.
- Passive systems must be designed with consideration for ease of ISI, testing and maintenance so that the dose to the worker is much less.
- Drawbacks

- Reliance on "low driving forces", as a source of uncertainty, and therefore need for T-H uncertainties modeling.

- Licensing requirement (open issue), since the reliability has to be incorporated within the licensing process of the reactor. For instance the PRA's should be reviewed to determine the level of uncertainty included in the models.

- Need for operational tests, so that dependence upon human factor can not be neglected.

- Time response: the promptness of the system intervention is relevant to the safety function accomplishment. It appears that the inception of the passive system operation, as the natural circulation, is conditional upon the actuation of some active components (as the return valve opening) and the onset of the conditions/mechanisms for natural circulation start-up

- Reliability and performance assessment in any case. Quantification of their functional reliability from normal power operation to transients including accidental conditions needs to be evaluated. Functional failure can happen if the boundary conditions deviate from the specified value on which the performance of the system depends.

- Ageing of passive systems must be considered for longer plant life; for example corrosion and deposits on heat exchanger surfaces could impair their function.

- Economics of advanced reactors with passive systems, although claimed to be cheaper, must be estimated especially for construction and decommissioning.

The question whether it is favourable to adopt passive systems in the design of a new reactor to accomplish safety functions is still to be debated and a common consensus has not yet been reached, about the quantification of safety and cost benefits which make nuclear power more competitive, from potential annual maintenance cost reductions to safety system response.

7. FINAL REMARKS

Based on the analysis of the critical aspects related to the open points discussed in the previous section a qualitative analysis, on the basis of the author's opinion, reported in Table 5 below aims at identifying for each of the above items both the criticality with respect to the passive system reliability assessment process, in terms of the relative importance and the existing advancement, according to Table 6 which ranks the relative level of both the importance and progress.

Table 5. Importance analysis

Item	Importance	Advance
Uncertainties	H	L
Dependencies	M	L
Integration within PSA	M	L
Passive vs. Active	H	L

Table 6. Grade rank for importance and advancement analysis

	Grade	Definition
Importance	H	The item is expected to have a significant impact on the system failure
	M	The item is expected to have a moderate impact on the system failure
	L	The item is expected to have only a small impact on the system failure
Advance	H	The issue is modelled in a detailed way with adequate validation
	M	The issue is represented by simple modelling based on experimental observations or results.
	L	The issue is not represented in the analysis or the models are too complex or inappropriate which indicates that the calculation results will have a high degree of ambiguity

It is clear that he worst case is characterized by "high "and "low" rankings relative respectively to the importance and the advancement aspects, thus making the correspondent item development a critical challenge.

Based on this, the results of this qualitative analysis show the relevance relative to the uncertainties and the comparison between active and passive, as most critical points to be addressed in the application of the PRA to the evaluation of the passive system performance assessment. This allows the analyst to track a viable R&D program to deal with these issues and limitations and to steer the relative efforts towards their implementation.

8. CONCLUSIONS

Due to the specificities of passive systems that utilize natural circulation (small driving force, large uncertainties in their performance, lack of data...), there is a strong need for the development and demonstration of consistent methodologies and approaches for evaluating their reliability. This is a crucial issue to be resolved for their extensive use in future nuclear power plants. Recently, the development of procedures suitable for establishing the performance of a passive system has been proposed: the unavailability of reference data makes troublesome the qualification of the achieved results. These procedures can be applied for evaluating the acceptability of a passive system, specifically when nuclear reactor safety considerations are important for comparing two different systems having the same mission and, with additional investigation, for evaluating the performance of an active and passive system on a common basis. The study while identifying limitations of the achieved results or specific significant aspects that have been overlooked has suggested areas for further development or improvements of the procedures:

- In order to get confidence in the achieved results, the reduction of the so identified level of uncertainty pertaining to the passive system behaviour, and regarding in particular the phenomenological uncertainty. In fact, it's worth noting that these uncertainties are mainly related to the state of knowledge about the studied object/phenomenon, i.e., they fall within the class of epistemic uncertainties, thus suitable for reduction by gathering and analyzing a relevant quantity of information and data.

- The determination of the dependencies among the relevant parameters adopted to analyze the system reliability.

- The study of the dynamical aspects of the system performance, because the inherent dynamic behaviour of the system to be characterized: this translates into the development of the dynamic event tree.

- The comparison against the active system, also to evaluate the economical competitiveness, while assuring the same level of safety.

Future research in nuclear safety addressing this specific topic relevant to advanced reactors should be steered towards all these points in order to foster and add credit to any proposed approach to address the issue and to facilitate the proposed methods endorsement by the scientific and technical community.

NOTES

[1] - In the following PSA (Probabilistic Safety Assessment) and PRA (Probabilistic Risk Assessment) are utilized indifferently

REFERENCES

1. IAEA TEC-DOC-626, 1991 Safety Related Terms for Advanced Nuclear Power Plants. September 1991.
2. IAEA TEC DOC-1474, 2005 Natural Circulation in Water Cooled Nuclear power Plants. Phenomena, models, and methodology for system reliability assessments, November 2005.
3. IAEA TECDOC-1624, 2009 Passive Safety Systems and Natural Circulation in Water Cooled Nuclear Power Plants. November 2009
4. United States Nuclear Regulatory Commission's (U.S. NRC) Reactor Safety Study (WASH-1400, 1975
5. Zio, E. , N. Pedroni, 2009 Building Confidence in the Reliability Assessment of Thermal hydraulic Passive Systems. Reliability Engineering and System Safety, 94 268281 .
6. Jafari, J. , F. D'Auria, et al. 2003 Reliability Evaluation of a Natural Circulation System. Nuclear Engineering and Design 224 79104 .
7. Marques, M. , L. Burgazzi, et al. 2005 Methodology for the Reliability Evaluation of a Passive System and its Integration into a Probabilistic Safety Assessment. Nuclear Engineering and Design 235 26122631 .
8. Lorenzo G., et al., Assessment of an Isolation Condenser of an Integral Reactor in View of Uncertainties in Engineering Parameters, Science and technology of Nuclear Installations, Volume 2011 Article ID 827354, 9 pages
9. Apostolakis, G. , L. Pagani, P. Hejzlar, 2005 The Impact of Uncertainties on the Performance of Passive Systems. Nuclear Technology 149 129140
10. Apostolakis, G. , F. Mackay, Hejzlar. and, 2008 Incorporating Reliability Analysis into the Design of Passive Cooling System with an Application to a Gas-Cooled Reactor. Nuclear Engineering & Design 238, 217-228

11. Burgazzi, L. , 2007a Addressing the Uncertainties related to Passive System Reliability. Progress in Nuclear Energy 49 93102 .

12. Burgazzi, L. , 2002 Passive System Reliability Analysis: a Study on the Isolation Condenser, Nuclear Technology 139 39 .

13. Burgazzi, L. , 2003 Reliability Evaluation of Passive Systems through Functional Reliability Assessment, Nuclear Technology 144 145151 .

14. Burgazzi, L. , 2007b Thermal-hydraulic Passive System reliability-based design approach, Reliability Engineering and System Safety 92 (9), 1250-1257.

15. Nayak, A. K. , et al. 2008 Passive System Reliability Analysis using the APSRA Methodology. Nuclear Engineering and Design 238 14301440 .

16. Burgazzi, L. , 2004 Evaluation of Uncertainties related to Passive Systems Performance. Nuclear Engineering and Design 230 93106 .

17. Ricotti, M. E. , E. Zio, F. D'Auria, G. Caruso, 2002 Reliability Methods for Passive Systems (RMPS) Study- Strategy and Results, in proceedings of the NEA CSNI/WGRISK Workshop on Passive System Reliability. A Challenge to Reliability Engineering and Licensing of Advanced Nuclear Power Plants, 146163

18. Burgazzi, L. , 2008a Reliability Prediction of Passive Systems based on Bivariate Probability Distributions, Nuclear Technology 161 17 .

19. Burgazzi, L. , 2009 Evaluation of the Dependencies related to Passive System Failure. Nuclear Engineering and Design 239 30483053

20. Burgazzi, L. , 2011 Reliability Prediction of Passive Systems with Multiple Degradation Measures, Nuclear Technology 173 153161 .

21. USNRC, 2007 Feasibility study for a risk-informed and performance-based regulatory structure for future plant licensing. US Nuclear Regulatory Commission, NUREG-1860.

22.]IAEA, 2007 Proposal for a technology-neutral safety approach for new designs. International Atomic Energy Agency, TECDOC-1570, Vienna.

23. Burgazzi, L. , 2008b Incorporation of Passive Systems within a PRA Framework. Proceedings of PSAM9, 9th International Probabilistic, Safety Assessment and Management Conference, Hong Kong, 1823 May 2008.

24. Mercurio, D., Podofillini, L., Zio, E., Identification and Classification of Dynamic Event Tree Scenarios via Possibilistic Clustering: Application to a Steam Generator Tube Rupture Event. Accident Analysis and Prevention 41 2009 1180-1191

25. Matzie, R. A. , A. Worrally, A. The, 1000 reactor-the Nuclear Renaissance Option. Nuclear Energy, 2004, 43, 1 Feb., 3345

26. Burgazzi, L. , 2008c About Time-variant Reliability Analysis with Reference to Passive Systems Assessment. Reliability Engineering and System Safety 93 16821688 .

27. JiYong Oh and Golay, M., 2008 Methods for Comparative Assessment of Active and Passive Safety Systems with respect to Reliability, Uncertainty, Economy and Flexibility. Proceedings of PSAM9, 9th International Probabilistic, Safety Assessment and Management Conference Hong Kong, 1823 May 2008.

Chapter 7

Policy analysis of China inland nuclear power plants' plan changes: from suspension to expansion

Jinxin Zhu[1] and Gail Krantzberg[2*]

1Institute for Energy, Environment and Sustainable Communities, University of Regina 3737 Wascana Parkway, Regina, SK S4S 0A2, Canada.

2Engineering and Public Policy in the School of Engineering Practice, McMaster University, 1280 Main St., W Hamilton, ON L8S 4 L7, Canada.

ABSTRACTS

Background

China's inland nuclear power plants plan has been suspended until 2015 since Fukushima disaster. The policy on inland nuclear power plants becomes uncertain. This paper provides an overview of inland nuclear power plants the safety grantee, economic power on diminishing disparities between Western China and Eastern China, efforts on environmental improvement with reforming of energy restructure and essential public participation. The paper further discusses the government's current policy and successful experience from other countries.

Results

The paper gives the recommendations for promoting inland nuclear power plants' expansion, making economy develop healthily, improving environment and introducing public hearing into the nuclear power development.

Conclusions

With proper political guiding, China could obtain significant benefits from expanding nuclear power plants from coast to inland.

KEYWORDS

Inland nuclear power plants, Public policy, West development strategy, Energy restructure

1. BACKGROUND

With China's rapidly developing economy in these years, the problems of power supply restriction, an energy bottleneck, oil and coal supply tension, and so on are emphasized. At the same time, large-scale use of conventional energy sources leads to increased environmental pollution especially in air pollution and GHS emission. In order to raise the proportion of clean energy and catch up with the changing trends, Chinese government makes its mind to accelerate the development of nuclear power as soon as possible.

Ambitious goals for NPPs expansion have been set by China government. By 2020, China must reach a 40-GW nuclear power generation capacity, have 18 GW of additional nuclear power capacity under construction, and ensure that nuclear power takes approximately 4% of electricity generation (i.e., 260–280 billion kWh) (Hou, et al. 2011). However, since Fukushima disaster in Japan, constructions of new nuclear power plants (NPPs)'s plan have been suspended from March 2011.

According to China's national nuclear plan, China will resume construction of new NPPs projects. Nevertheless, it had lowered its target for the construction of NPPs by 2015, notably by not building additional nuclear reactors at inland locations. Nuclear Power Safety Plan (2011-20) as well as the Mid-term and Long-term Development Plan for Nuclear Power (2011-20) was discussed and passed through executive meeting on Oct 24th, 2012. During the 2011-2015 periods, not any nuclear projects will be constructed in inland regions, but only a few projects in coastal areas that have gone through justification processes will be constructed (Xinhua 2012).

While acknowledging that nuclear power expansion plan in China presently is exhibiting symptoms of extreme stress from a combination of sources that include safety concern, economy concern, and environment concern. Moreover, public concern of NPPs' safety is not ignoring due to Fukushima incident. This paper is going to analyze factors related to China inland NPPs' strategies and discuss the current government policies, potential changes to current policy from 2012 to 2015, the future of nuclear power after 2015, and the barriers to nuclear power development in China. Some recommendations are also provided.

Inland location NPPs from developed countries and China's high temperature Gas cool reactor technology (safety concern)

Other countries' primary considerations for NPPs construction are economic and social factors like local demand of electricity, and the situation of the primary energy and power distribution, there is not a distinction in safety between coastal and inland NPPs in the site selection. As a truth, countries like the Canada, France and United States all build a lot of NPPs in inland areas, and the amount of generating reactors and total installed capacity are much higher than coastal NPPs (Dynabond Powertech 2010). General information of percentage of inland reactors is shown in Table 1.

Table 1. General information of world's inland reactors(National Energy Administration **2012**)

Nations	Number of inland reactors	Number of total reactors	Percentage of inland reactors
Global	220	442	49.8%
USA	64	104	61.5%
France	40	58	69.0%
Canada	12	14	85.7%
Germany	14	17	82.4%
Russia	18	31	58.1%
Ukraine	19	19	100%
China	0	11	0%
Others	53	199	26.6%

From regulations on NPPs of International Atomic Energy Agency and main nuclear reactors possessing countries, the security objectives and evaluations criteria on coastal nuclear power stations and inland NPPs are identical. There is no difference between regulations of inland NPPs and coastal NPPs' developments. No countries and organizations had ever post special requirements on inland nuclear power plants' safety (National Energy Administration 2012). These countries' operating experience of NPPs demonstrates that safety of inland NPPs can be fully guaranteed.

From Table 1, of the total 442 nuclear stations around the globe, 50 percent are located inland. In Canada, that number is at a striking 86 percent.

After Fukushima disaster, safety of traditional nuclear power plant technology is in fierce debate all over the country, construction work on a power plant using domestically developed, fourth generation technology has already began in Rongcheng City, Shandong Province in 2012. Because the project gained approved prior to the State Council's decision to impose a moratorium, it were unaffected by suspension of new NPPs project approvals VelkerMatt (2011). Fourth generation reactors represented by high temperate gas-cooled reactors are main trend of NPPs development in China.

China developed the fourth generation technology entirely. In 2003, China's first experimental modulized pebble bed high temperature gas-cooled reactor plant was designed, built by Tsinghua University and began to operate and achieved full capacity operating and supplying power with 10,000-kilowatt to the grid. In terms of its safety, the fourth generation is obviously superior to the former generations' technology, because it uses helium coolant's natural convection to control the reactor even in high temperatures (NGNP Industry Alliance 2012). The pressured water reactors units we see today are massive. However, the actual reactor itself is only about one thousandth the size of the total unit in reality. With the reactor made sufficiently large, it can radiate heat more easily, meaning no power and no water or other cooling fluid is required for high temperate gas-cooled reactor. If normal heat transport systems are not available, heat removal from the reactor occurs naturally and directly to the earth. Even if normal cooling systems are not functional, it takes days to reach limited temperatures (still under temperatures that could melt reactor core) with low energy density of the reactor core and the large heat capacity of the graphite structure. Therefore, reactors can be built and operated independent of water source.

2. RESULTS AND DISCUSSION

2.1 Economy concern

For planning purposes, China's 31 provinces are divided into three macro-regions–eastern, central and western(see Figure 1). The developments of three regions are ranked with respect to the greatest in the first and the last in the third. Two kinds of disparities between Western China and Eastern China pervade China. The first one is regional disparities that focus on levels of economic development and it varies considerably from Western China to Eastern China. The second is referring to mismatch in energy resources supply and demand. Eastern China suffers severe shortages while Western China is blessed with significant surpluses (HuangTodd and ZhangDaniel 2011).

Figure 1. Three macro-regions of China (HuangTodd and ZhangDaniel **2011**).

One important indicator, Vw coefficient, confirms the existence of regional disparities. The coefficient is used by Williamson (1965). With the regions being represented by China's provinces, it is a measure of variation in weighted GDP per capita the proxy for development–across a country's set of regions. A Vw of zero means the ideal outcome of an absence of any regional departures from a national average GDP; a coefficient of unity represents that the localization of wealth in a single region while all others were destitute. A Vw analysis fortheperiod 1952–2008(with per capita GDP standardized on the former year, see Figure 2) demonstrates that Chinese regional disparities have increased shapely and consistently exceeded 0.7 since 1970 (HuangTodd and ZhangDaniel 2011). Little could be detected in the way of gap closing for much of that period, but some trend towards reduction in disparities has become discernible since the West Development Strategy initiated.

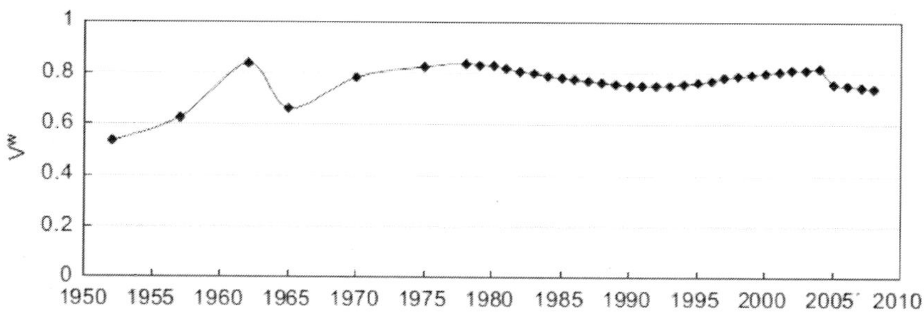

Figure 2. Chinese regional disparities measured by Vw coefficients, 1952–2008 (HuangTodd and ZhangDaniel**2011**).

This disparities problem not only causes the economic issues to Western China, but also makes the environmental and social issues more severe in that region. Due to the underdeveloped economy, Western China is too dependent on mining energy or non-energy minerals and the over mining makes the fragile ecosystem even worse. Its education and welfare stay in quiet low level for decades because of lack of investment. These issues deter the region's economic development in reverse. In national scale, disparities between Western China and Eastern China in China threat its well-being and integral development. To balance the development of Western China and Eastern China, China government established Western China Development Strategy in 1999 (HuangTodd and ZhangDaniel 2011). This strategy are mainly focusing on pulling demand and market of Western China, reforming the energy structure which is going to be discussed in environmental concern to conserve the ecosystem in this region, keep the society stable, and secure the borderland of China.

Nuclear plant and its construction will boost local economies because of its significant economic benefits. From direct and secondary effects, the average nuclear plant produces approximately $470 million economic output in the local community and nearly $40 million in total labor income for each year

(Nuclear Energy Institute of United States 2012). The direct effects reflect the labor, goods and services of plant's expenditures. The secondary effects include subsequent plant expenditures related to the presence of the plant and its employees that penetrate into the local economy. Analysis illustrates that $0.04 net benefits in the local community come from every dollar spent by the normal NPPs' construction. The NPPs generates nearly $16 million in local tax revenue for each year (Nuclear Energy Institute of United States 2012). The tax revenue could benefit education, transportation, and other local infrastructure. Nuclear power plants provide substantial economic benefits during their decades of operation. The taxes, direct and secondary expenditure from NPPs strengthen the economies of local communities and hosting a new nuclear plant will give a similar boost to any community. With different plant size, a new nuclear plant represents an investment ranging from $6 billion to $8 billion including interest (Nuclear Energy Institute of United States 2012). Moreover, construction of a new nuclear power plant will provide a substantial boost to manufacturers and suppliers of commodities such as steel, concrete and cement.

Current Chinese nuclear expansion plan that only covers the coastal regions is going to aggravate the disparities between these two regions. For NPPs' strong power on boosting economy, they can be a powerful tool to cooperate with Western China Development Strategy to realize its main purposes.

2.2 Restructure of energy (Environment Concern)

In China, coal covering 60% of the primary energy and 75% of the power generation leads to irrational energy structure. The coal based energy structure causes serious problems of environmental pollution and emission of greenhouse gases (GHG). China has become one of the most polluted countries in the world. For example, China emitted 20 Mt of $SO2$ and 3 Gt of $CO2$, both ranked the world No. 2 in 2006. Environmental pollutions cause the economic loss accounting for 3–7% of GDP in China. If these problems could not be solved properly and quickly, they will hinder the further development of the country. The China Energy Policy (2012) published by the State Council said that China decided to make up to 45 percent reduction of its carbon dioxide emissions by 2020 (Tiantian 2012). To achieve that goal, China has to shut down numerous coal plants and replace them with clean energy plants.

Simultaneously, compared the consumption of energy with east which is always in energy crisis, west only took less than 30% of total. As shown in Figure 3.

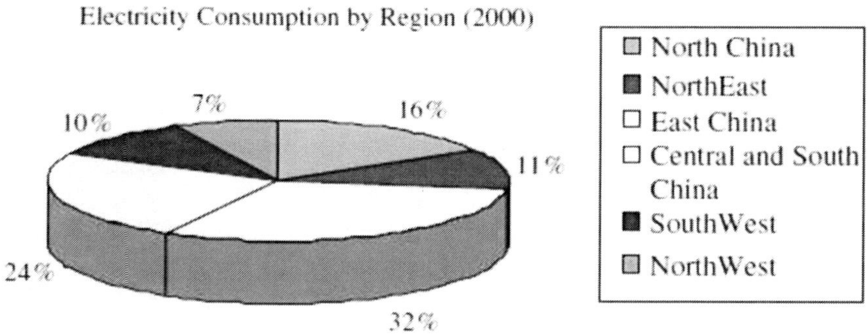

Figure 3. Electricity consumption by region (2000) (China Statistics Press **2002**).

The West-East Power Transmission Project aims to turn the rich energy resources in West China into electricity and to transmit the electricity to East China, the economically dynamic and energy-deprived region of the country (Ouyang, and Changyu 2002). It is an important part of the West Development Strategy of the Chinese government, which will enhance the optimization of energy resources contributing to the healthy development of the national economy. This project will not only solve the energy crisis and the severe environmental contamination of Eastern China but will also transform the abundant resource of Western China into an economic advantage that will help mitigate poverty there. In this way, the West-East Power Transmission Project will promote the sustainable development of the whole country.

There are three main national power flows from Western China to the southeast coastal areas, in particular, the Guangdong province and Shanghai city referring to the West-East Power Transmission Project (Kang et al. 2012). As shown in Figure 4.

As a result of the rapid expansion of the West-East Power Transmission Project, air pollution is serious in the northern passageway region mainly where consists of coal-fired power plants, especially for Eastern China, and the main air pollutants are PM and SO2 (Zhigang et al. 2005). Thus, the government and public have shown great concern about the proposed power plants in this region's impacts on environment. It is very important, therefore, to understand the impacts of building new clean power plants on improving the air quality in this region.

Nevertheless, Western China's energy shortfalls have surfaced with respect to remaining reserves. China's total remaining oil reserves 1.48 billion barrels by the end of 2009. The reserves/production (R/P) ratio of it was 10.7 years, which was far below the world's average oil R/P ratio of 45.7 years. In terms of remaining natural gas, it reserves totalled 2.46 trillion cubic meters with an R/P ratio of 28.8 years, which is lower than the world's average natural gas R/P ratio of 62.8 years. Remaining coal reserves 1.145 billion ton in total with an R/P ratio of 38 years which is far below the world's average of 119 years (British Petroleum (BP). 2011). Western China's energy shortfalls will not only affect local development, but also will threat the whole country's economy growing and environment improving. Needs for clean energy plants planned in these regions are urgent now.

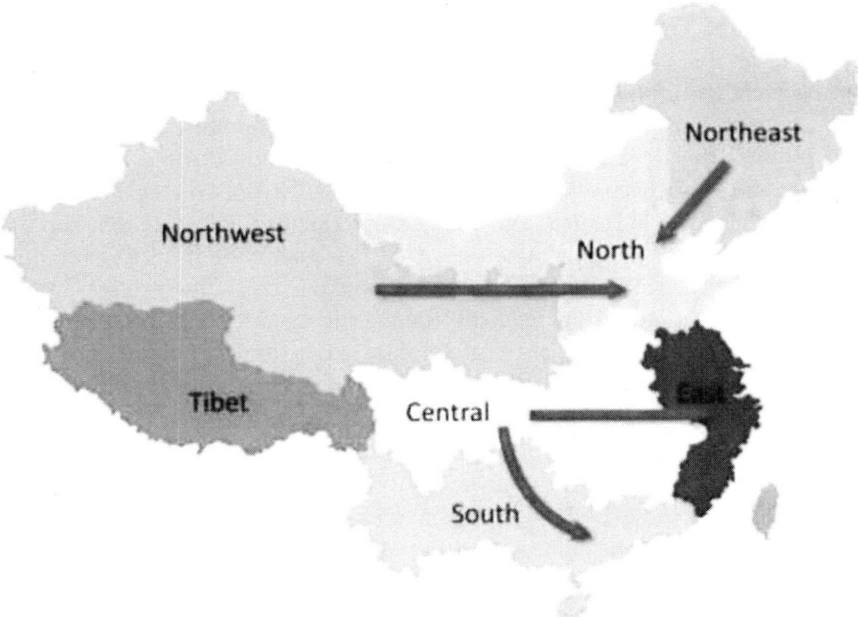

Figure 4. China's national power flow pattern (Ouyang and Changyu **2002**).

2.3 Public concern

In addition to technical and economic factors, the public understanding and acceptance to the nuclear power play an important role in the development of nuclear power. The familiarity of public towards nuclear power that is terminology of public understanding is an important impact factor on public acceptance. With the increased familiarity with nuclear power, the proportion of people who consider NPPs' safety was corresponding increased. The percentages of public, scientists, energy scientists and nuclear experts who believed the NPPs' safety were respectively 40%, 60%, 76% and 99% (Barke RP HC Jenkins-Smith 1993). A study on the level of knowledge and attitude of nuclear power among Chinese residents around the Tianwan nuclear power plant within a radius of 30 km in Jiangsu of China shows the level of understanding and acceptance of nuclear power was generally not high (Ningle et al. 2012). Other 10 operating NPPs in China have the same situation.

Low level of understanding and acceptance of nuclear power from public will cause serious civic activism like local protest groups. Main issue prompting some local groups and communities to oppose nuclear projects is sensitivities over nuclear fuel leak risks. Local protest groups and adverse public opinion are emerged to challenge the legitimacy of proposed projects by a strong NIMBY (not-in-my-back-yard) element, especially in rural areas (International Atomic Energy Agency 2002). Although Chinese history is rife with prominent examples of civic activism, opposition groups have been unable to stop nuclear power development. When a small group from local communities tried to protest licensing of the Daya Bay plant, officials' detained and arrested them. When more than one million people in Beijing signed a petition against nuclear power, Chiang Hsin-Hsiung, the Minister of Nuclear Industry, rebuffed the complainants by issuing a statement that the government "the unscientific objections from some people would not halt the Daya Bay project". When United Kingdom authorities and the Hong Kong Civic Association came out a series of reports questioning the safety of Daya Bay, the government answered "reports compiled by other countries do not constitute legal documents" and thus held no sway over the licensing process (Zhou SXZhang 2010). With no nuclear power mishaps happening, civic activism against nuclear power has waned since the late 1980s. However, the recent events at the Fukushima Daiichi nuclear power plant in Japan have resulted in a great focus on understanding both the safety of nuclear power and the public understanding and acceptance of nuclear power (FoE, F.o.t.E.,GfKNOP 2011). Coercion policies of Chinese government like these, nowadays, lose their power on controlling of society and result in serious potential problem for social security.

On August and October 2012, plans to expand petrochemical plants in northern and eastern China have been shelved after days of protest (BBC NEWS 2012). The plants are producers of PX (a carcinogenic petrochemical used to create raw materials for the production of polyester film and fabric). Tens of thousands protested against chemical plants in Dalian city and Ningbo city. Environmental protests have been more common and influential in China. Many Chinese are becoming more environmentally aware and are deeply concerned about pollution and living quality surrounding them. China policy makers started to pay attention to public voice on NPPs. There are some changes made in its official documents. On March 2012, the National Nuclear Safety Administration (NNSA) announced "if the environmental impact report does not have a public participation chapter, NNSA would not accept the report. According to the announcement of China's State Environmental Protection Department (SEPA), public participation will run through the whole process, and projects which have strong objections are likely to be stopped." (Dynabond Powertech 2010).

To deal with public issues, experience from other countries is quiet valuable for China. The UK Government's latest National Policy Statements for Energy Infrastructure recognised "there may be positive and negative effects from the construction of new nuclear plants and that organisations applying for site license should identify these effects in their license application." (DECC 2010). Although there are no specific comments on mitigating or avoiding these effects, the activity involved public participation is still widely adopted in nuclear power area in UK. The UK Health and Safety Executive has invited public input into its Generic Design Approval (GDA) process for the assessment of new nuclear reactor designs (HSE 2009). However, as the designing or licensing is almost finalised, any changes are might very costly and time consuming, and it is rare that such changes that are not critical for the safety will be made, even if they have potential ability to enhance the socio-economic benefit of the plant. Therefore, making changes earlier in the designing and licensing process would be less costly and more efficient.

3. METHODS

After Fukushima disaster in Japan, China NPPs' expansion map changed as shown in Figure 5. Only coastal NPPs passed the new nuclear power plant project approvals. This is because dumping radioactive contaminations into sea has much lower risk and cost than dumping into river once the leaking happened in NPPs. However, only selecting locations for NPPs in coastal areas will not be able to meet the future needs of China's sustainable development. To maintain the energy supply for inland provinces which are suffering energy shortfalls and environmental contamination, NPPs are urgently need to be constructed and planned in these regions (Dynabond Powertech 2010). Aforementioned safety concern is the premise for satisfying this urgent need. The successful development of inland NPPs in several leading country such as USA, France and Canada shows that safety is not a main barrier for China to overcome to cover its inland province with nuclear power. Besides, the maturity of China self-developed high temperature gas cool reactor technology will make the siting process of NPPs independent from the limitation of closing to shore of river, lake or coast. Government has large flexibility to cover the benefits of NPPs expansion plan with inland provinces.

But the Fukushima disaster does reveal some designing flaws in NPPs like insufficient emergency producers design. So, revaluation and reexamination is necessary to the initial design of inland NPPs during the 3 years suspension from 2012 to 2015. However, if inland NPPs construction is put off easily by a slight risk, it would be a huge loss of great opportunities to achieve sustainable economy for China.

Figure 5. Power reactors in mainland China and on Taiwan (World Nuclear Association **2012**).

Disparity between west and east in China hinder the economy from developing healthily. Primary barrier is energy and resource misplacing as mentioned in previous section. West has main fossil fuel resources storage in China.

Misplacing and energy shortfalls make Western China's economy much more depressed and make disparity from Eastern China more severe. Following Western China Development Strategy, NPPs should be planned at inland locations. Not only boost local economy at west, but also improve the mobility of factors of production between west and east. The economic and trading cooperation and the mobilization of capital between Western China and eastern regions are the productivity of capital and could complement technological advances in Western China of the country. With these benefits, construction and planning of inland NPPs will contribute to diminish disparities and eliminate the risk of the economic gap between Western China and Eastern China being lager. Therefore, improve the society stability in Western China and remit the contradictions in whole country.

Western China-East Power Transmission Project has been proved to make a great contribution to environment improvement of China by reducing the transmission and burning of fossil fuel. However, coal fired plants from northern passageway of the project's environmental problems are still serious and these problem have the trend to be worse due to the big energy gap for both west and east. To mediate the energy supplying intensity and improve the air-quality in western regions, connecting clean energy generators to electricity grids of northern passageway would be dispensable. In perspective of Chinese electricity capacity development, percentage of coal-fire electricity will decrease with respecting the growing percentage of nuclear power (see Figure 6). In terms of nuclear power's consistency supplying and low cost to generate characteristics, it is most stable and affordable power resource for west compared with solar and wind power. Policy makers should consider increasing the nuclear power's percentage of total energy transmission.

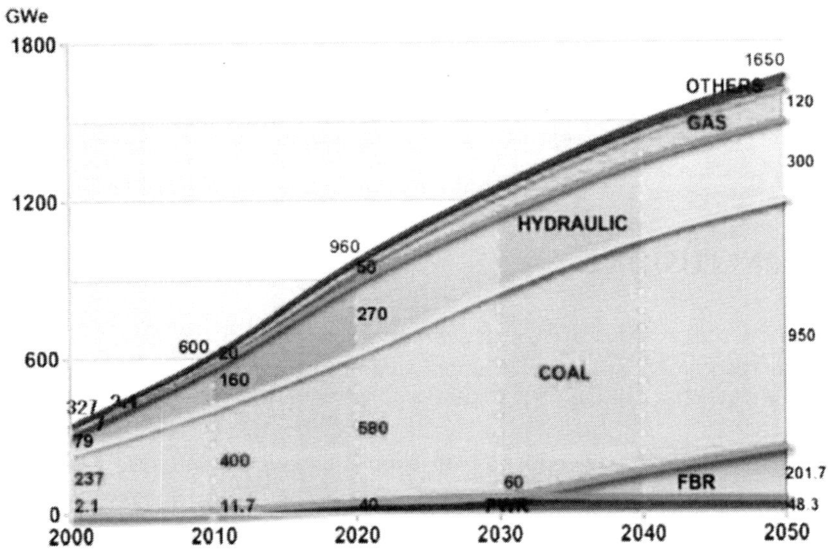

Figure 6. The electricity capacity development envisaged in China by 2050.

By involving the public at an earlier stage than is currently common, a nuclear plant can be built on a foundation of social consensus from most leading countries. Currently, interactions between the public and nuclear industry start with the licensing process and continue through to decommissioning in English. By extending this interaction to earlier stages, specifically during requirements analysis and plant design, it ought to be possible to increase public understanding of nuclear power and minimise the probability for disruption to new-build activities caused by misunderstanding and mistrust. It may not be possible to integrate all of the public views into a nuclear plant design, however the process of dialogue involved in such discussions can only reinforce current educational campaigns and help in building mutual trust. It is essential for keeping the social in Western China stable and the premise for promoting the inland NPPs' expansion.

For public participation, the operating and under constructing NPPs in eastern regions are all missing in their designing and licensing processes. The proposed NPPs projects in eastern regions are facing complicated public problems integrated with their affecting sources included the high population intensity, diverse education background and family income. In term of western regions' public problems, they are simpler than Eastern China's because communities in Western China has low population density, similar education background (average is quiet low) and not big difference in family income. To deal with public issues, west would be amore ideal test field. Having the experience and experimental data for public participation from west, next step is to apply them to future proposed projects especially located at east where has much more intricate situations. The process is like going from easy to hard. All stakeholders easy accept progressive reform.

4. CONCLUSIONS

Based on the analysis in this paper and experiences from other countries, several recommendations to promote inland nuclear power development in China are given as follows:

- Inland Nuclear Power Plant plan should be written into Western China Development Strategy. After 2015, inland NPPs construction should be resumed. Under Western China-East Power Transmission Project, the electricity that is generated from the starting point of northern passageway grid must have no less than 5% from NPPs by 2020.
- During 3 years suspension from 2012 to 2015, we should attach great importance to the public acceptance of NPPs, enhance communication

with the local government and local people by holding public hearing and meeting. In early designing stage, government is responsible to hold a public hearing for enhancing nuclear safety awareness, eliminating people's concerns, and creating a good social environment for the development and construction of nuclear power projects. In licensing stage, individuals who are directly affected by any licensing action involving a facility producing or utilizing nuclear materials have rights to ask for public hearing. Central government should give legislation support for all three hearings by establishing the formal regulations and local governments will be in charge of arranging the time and places to hold hearings.

- From 2015 on, all proposed NPPs projects will be evaluated their public participation chapter through early designing and licensing process by China's State Environmental Protection Department (SEPA). Reports without or forged public part will be rejected immediately. Public participation will run through the whole process and projects that have strong objections must be stopped.

AUTHORS' INFORMATION

ZJ's background is civil engineering. He received his master degree of engineering and public policy in Sept, 2013. Now, he is a PhD student under the supervision of Dr. Gordon Huang in University of Regina. He is performing as a research assistant in the Institute for Energy, Environment and Sustainable Communities.

KG is the Professor and Director of the Centre for Engineering and Public Policy in the School of Engineering Practice in McMaster University. KG was the direct supervisor of ZJ from 2012 to 2013.

ACKNOWLEDGEMENTS

We would like to thank the individuals who have given their time and expertise to in reviewing and offering advises as follows: Huijing Xu (Shanghai, China), Charles Jin (Hamilton, Canada) and Younggew Kim (Seoul, Korea).

COMPETING INTERESTS

The authors declare that they have no competing interests.

AUTHORS' CONTRIBUTIONS

ZJ designed the study, collected and interpreted data, conducted data analysis and wrote the manuscript. KG revised the manuscript critically and approved the final version to be published. Both authors read and approved the final manuscripts.

REFERENCES

1. Barke RP, Jenkins-Smith HC: *Politics and scientific expertise: scientists, risk perception, and nuclear waste policy*. Risk Analysis; 1993:425–439. McLean McLean
2. BBC NEWS: *China Protesters Force Halt to Zhejiang Factory Plan*. 2012. http://www.bbc.co.uk
3. British Petroleum (BP): *Statistical review of world energy 2010*. 2011. http://www.bp.com/statisticalreview
4. China Statistics Press: *China Statistical Yearbook 2002*. 2002. http://www.stats.gov.cn
5. DECC: *Consultation on the revised draft NPSs for Energy Infrastructure*. London; 2010.
6. FoE, F.o.t.E.,GfKNOP: *Post Fukushima public perception of nuclear power*. Japan: Energy of Japan; 2011:42–50.
7. Hou J, Zhongfu T, Jianhui W, Pinjie X: **Government policy and future projection.** *J Energy Eng* 2011, **137:**151–158.
8. HSE: *New nuclear power stations, Generic Design Assessment, Safety assessment in an international context*. Merseyside: Health and Safety Executive; 2009.
9. HuangTodd Y, ZhangDaniel L: **Capitalizing on energy supply: Western China's opportunity for development.** *Resour Policy* 2011, **36:** 227–237.

10. International Atomic Energy Agency: *People's Republic of China. country profiles*. 2002, 214–235. http://www.iaea.org

11. Kang C, Zhou T, Chen Q, Qianyao X, Xia Q, Ji Z: **Carbon emission flow in networks.** *Sci Rep* 2012, 34–45. Beijing Beijing

12. National Energy Administration: *Discussion on Safty of China Inland NPPs*. Beijing: National Ennergy Administration Energy Science; 2012.

13. NGNP Industry Alliance: *The High Temperature Gas-Cooled Reactor (HTGR) - Safe, Clean and Sustainable Energy for the Future*. High Tempreture Gas Rectors; 2012:11–15.

14. Ningle Y, Zhang Y, Wang J, Cao X, Fan X, Xiaosan X, Wang F: **Knowledge of and attitude to nuclear power among residents around Tianwan nuclear power plant in jiangsu of China.** *Int J Med Sci* 2012, 361–369.

15. Nuclear Energy Institute of United States: *Nuclear Power Plants Contributes Significantly to State and Local Economies*. Reliable and Affordable Energy; 2012.

16. Ouyang, and Changyu: *Implement the "Power Delivery from the West to East" strategy to promote the economy development of East and West China*. Beijing: Electric Power; 2002:11–15. in Chinese in Chinese

17. Powertech D: *Study on China's inland NPPs*. Dynabond Powertech Service; 2010.

18. Tiantian B: *Nuclear coastline*. Global Times; 2012.

19. VelkerMatt: *Construction of China's 4G nuclear reactor to start soon*. 2011. http://www.china.org.cn

20. World Nuclear Association: *Nuclear Power in China*. 2012. http://www.world-nuclear.org/info/inf63.html

21. Williamson, Jeffrey G: **Regional Inequality and the Process of National Development: A Description of the Patterns.** *Economic Development and Cultural Change* 1965, **no. 4 (Julho):** 1–84.

22. Xinhua: *No inland nuclear power plants to be built during 2011–15: statement*. Beijing: People's Daily Online; 2012.

23. Zhigang XH, Fahe C, Ning D, Yizhen C, Jindan L, Fu C, Simei L, Wenqing PJ: **Air quality impact of the coal-fired power plants in the northern**

passageway of the china west-east power transmission project. *J Air Waste Manage Assoc* 2005,**55**(12):1816–1826.

24. Zhou SXZhang: *Nuclear energy development in China: a study of opportunities and challenges. Energy*. Shanghai: Energy 2010; 2010. in press in press

Chapter 8

Selected Chemical Aspects of Nuclear Power Development

Andrzej G. Chmielewski, Marcin Brykala, Tomasz Smolinski

Institute of Nuclear Chemistry and Technology, Warsaw, Poland

ABSTRACT

The Fukushima nuclear power plant (NPP) accident consequences are a new challenge for nuclear power development; however the sequence of the event has illustrated importance of radiationand radiochemistry processes on the safe operation and shut down of nuclear reactor and decontamination of formed liquid and solid wastes. A chemistry program is essential for the safe operation of a nuclear power plant. It ensures the integrity, reliability and availability of the main plant structures, systems and components important to safety, in accordance with the assumptions and intent of the design. The proper implementation of these procedures minimizes the harmful effects of chemical impurities and corrosion on plant structures, systems and components. It supports the minimization of buildup of radioactive material and occupational radiation exposure as well as limiting of the release of chemicals and radioactive material to the environment [1].

KEYWORDS

Chemistry; Nuclear Safety; Water Radiolysis; Radioactive Waste Treatment

1. INTRODUCTION

The history of radiochemistry has started over hundred years ago. The first separation scheme was developed by Maria Skłodowska-Curie and Pierre Curie [2]. The biggest development of the science was related to military and peaceful applications of fissile and radioactive elements in nuclear power, industry and medicine. The role of chemistry in the present safe and effective operation of all stages of nuclear power related technology is unquestionable.

The Polish Government decided to lunch a Polish Nuclear Energy Program which assumes the construction NPP by the year 2023. The Institute of Nuclear Chemistry and Technology (INCT, known in Poland as IChTJ) have elaborated own programme considering a role of their expertise in radiochemistry, radiation chemistry, radiobiology and radiometric methods to support of this programme. The construction of the Center of Radiochemistry and Nuclear Chemistry for Nuclear Power and Nuclear Medicine has been completed this year.

2. CHEMISTRY AND SAFETY

2.1. Radiomonitoring, Radioecology and Emergency Preparedness

Monitoring of fuel clad integrity, the first and the most important barrier against radioactivity release into the environment, is the most important line of process radiation monitoring at nuclear power installations (NPIs). Current radioanalytical methods for detection of such radionuclides are usually difficult and time consuming, they require a great effort for sample separation and long-time measurements. Inductively coupled plasma mass spectrometry (ICP-MS) can be a complementary method to radiometric ones. ICP-MS methods of radionuclide determination have started to be developed since the last 15 years. In the INCT works on ICP-MS methods for U, Pu, ^{90}Sr, ^{241}Am are being carried out. The procedures are developed mainly for application to environmental analysis and preparation of test materials for proficiency tests on radionuclide determination which the INCT is a provider [3].

2.2. Coolant Chemistry, Reactor Elements Corrosion and Decontamination of the Primary Circuit

Coolant chemistry and reactor elements corrosion plays a very important role in nuclear engineering, covering all aspects of water radiolysis in boiling water reactor (BWR) and pressurized water reactor (PWR) units, effects of radiolytic phenomena on corrosion, etc.

There are three main tools used to handle reactor coolant system chemistry: pH control, zinc injection and dissolved hydrogen [4].

Decrease in pH of the water causes corrosion of metal equipment and destruction of water supply. Optimal pH of primary water approximately 7.2 +/– 0.2 is recommended, because that minimizing the solubility and corrosion of the surfaces of the main components in the primary system, the difference of solubility and thus of transport of radioactivated isotopes in the various parts of the system at different temperatures, typically in the 270°C - 325°C range. In practice, if the duration of a fuel cycle is extended 18 months or more is extending, it meant that the boron concentration at the beginning of cycle had to be raised from 1200 to 1800 mg/kg (ppm). This requires a lithium concentration greater than 2.2 mg/kg to maintain optimum pH.

The correct pH of water is very important for minimum solubility of nickel ferrite, which is the major component of CRUD. It is product of corrosion, mainly R_3O_4 (R = Fe, Ni or Cr), and a ratio of these elements depends on the type of construction material and type of reactor. Crud circulates inside the cooling system and in the vicinity of the core is activated, then settles on poorly secured locations of pipes, cooling system components and etc., which is a major threat.

The next tool used to handle reactor coolant system chemistry is addition of zinc, where is used zinc's affinity for the tetrahedral sites of spinel oxide minerals' structure. Thus, it can displace other cations such as iron, or nickel from oxide films, which activates into ^{58}Co, a major radioelement. Zinc is added at a concentration of about 5 μg/kg (ppb), which is sufficient to reduce the dose rates by 15% during the first fuel cycle and up to 30% to 40% after a few cycles with zinc injection. Relatively high cost of zinc depleted (^{64}Zn) salts causes searching new additives like aluminum salts. A study by gel chromatography in combination with radioisotope and elemental analyses showed that the Co(II) ions incorporated in polymeric hydrolysis products in the step of their formation were practically fully replaced by aluminum ions in the course of the polymer aging. Because such polymers are precursors of loose corrosion product deposits formed in primary circuits of NPIs and incorporating the most radiation-hazardous radionuclide ^{60}Co, addition of aluminum salts to primary coolants improves the radiation situation in the maintenance area of NPIs and, therefore, can be appropriate solution [5].

The third important tool to handle reactor coolant system chemistry is to optimize hydrogen concentration, applied to mitigate water radiolysis. For a long time, a value of 25 - 35 ml/kg (within a range of 25 - 50) has been used. The US Electric Power Research Institute (EPRI) is considering increasing the hydrogen concentration towards 50 ml/kg and possibly beyond, depending on the results of the ongoing qualification work on the risk of decreasing the time to crack initiation. The intention is to minimize crack growth rates. However, Japanese utilities are evaluating the action of decreasing hydrogen concentration to delay crack initiation.

On the other hand, the knowledge of radiolytic processes governed by physics and chemistry plays a very important role in the development of fuel reprocessing technologies, waste reprocessing and spent fuel and conditioned waste storage.

One of the challenges of chemistry is decontamination coolant water. It is necessary to establish precise chemistry to achieve optimum results. There are a lot of chemical and physical processes used to decontaminate coolant. It is used ion exchangers, solvent extraction and membrane methods. Water in primary coolant circle might be contaminated by fission products for example cesium or strontium. It might be removed from coolant by using special sorbents. In INCT by J. Narbutt groups synthetized sorbent to composite ion exchanger for selective removal radiocesium [6]. The sorbent allow to remove cesium ions from aqueous solutions containing cations of other metals.

2.3. Fuel Reprocessing

Generation IV reactor concepts and accompanying reprocessing technologies aqueous-based and pyrometallurgical, is a new challenge from the chemistry point of view [7]. Mainly, how to reduce the cost and to improve the safety of the management of such high-level nuclear waste are being considered.

An advanced closed fuel cycle relies on the possibility to maximise the energy usage of nuclear spent fuel and to provide improved waste forms for long-term storage: the removal of minor actinides (i.e.: Np, Am, Cm, Cf) with U and Pu from the waste and in some cases long-lived fission products (LLFP: ^{99}Tc, ^{129}I and ^{135}Cs) and heat generation fission products (^{90}Sr and ^{137}Cs), are partitioned by chemical separation. Separated all elements of significant Trans-Uranium (TRU) elements are immobilizated or transmutated in either a thermal or in fast neutron reactors [8] into short-lived or stable isotopes prior to disposal.

Various technologies of partitioning, based on hydrometallurgical processes, were developed to extract the minor actinides from nuclear waste—an acidic (HNO_3) solution as a rule, the aqueous raffinate, which remains after recovering uranium and plutonium in the solvent extraction process PUREX [9]. The technique can also be extended for the recovery of neptunium, but separation of trivalent americium and curium from trivalent lanthanides (Ln) cannot be separated directly in this process. Studies on solving of this issue are carried out for decades by many research teams.

Therefore, two steps of separations processes are at present considered. In the first step, at high acidity, a group separation of MA and lanthanides is carried out, followed by a selective separation of MA from lanthanides at lower acidity.

In this technology developed by the EU consortium coordinated by C. Madic, in the first step the minor actinides are directly extracted from the PUREX raffinate together with fission lanthanides as nitrates, using diamide extractants—the malonamides or diglycolamides (DIAMide EXtraction process) [10]. Various extraction systems were studied for the second step. The MA (III) + Ln (III) mixture generated after this first step is lowacidic to facilitate the second process, the SANEX process (Selective ActiNide EXtraction), which the goal is to separate the trivalent An from the Ln directly from the DIAMEX product. This process is based on the BTP, which belongs to a new family of extractants, the BisTriazinyl-Pyridine developed by Z. Kolarik et al., and is very efficient for a selective extraction of MA (III) at high acidity [11] and shows good capabilities in centrifugal extractors.

The EU ACSEPT project in which INCT is active plays a very important role to achieve the progress in the chemical separations field [12]. One of the concept of ACSEPT project is optimize of GANEX (Group ActiNide EXtraction) In the second cycle of separation of actinides from lanthanides is possible, as well as from other fission products. Objective is achieved by using as the organic phase of extraction mixture of bis-triazinebipyridine (BTBP) and tri-n-butyl phosphate (TBP) in cyclohexanone. The aqueous phase contains 4 M nitric acid, actinides, and lanthanides, and FP. BTBP molecules containing soft-donor nitrogen atoms favoring complexation of actinide 5f orbitals have the ability to separate actinides (III, IV, V and VI) from lanthanides (III). The TBP allows the extraction of uranium and plutonium [13].

One promising concept after the partitioning step is embedding the minor actinides (MA: Am, Cm, Cf) in uranium-based nuclear fuel by sol-gel process. This will allow the MAs to be destroyed by fast-neutron reactions in the upcoming generation-IV reactors.

Even a manufacturing of uranium oxide fuels is well developed process, synthesis of mixed uranium-MA fuels is a process under development. At the

INCT a new variant of a sol-gel method called—Complex Sol-Gel Process (CSGP) has been elaborated to obtain uranium dioxides (Patent Pending—[14]). This method has been used to synthesis of uranium dioxides doped by neodymium, as surrogates of trivalent plutonium and americium. The main modification step is the formation of uranylneodymium-ascorbate sols from components alkalized by aqueous ammonia. Those sols were gelled into: 1) irregularly agglomerates by evaporation of water; 2) medium sized microspheres (diameter < 100 μm) by INCT variant of sol-gel processes by water extraction from drops of emulsion sols in 2-ethylhexanol-1 by this solvent (**Figure 1**). Uranium dioxide was obtained by a reduction of gels with gas mixture of argon and hydrogen at temperatures 900°C [15]. The work has been carried out within the collaborative Project ACSEPT (Actinide Recycling by Separation and Transmutation), contract No. FP7-CP- 2007-211267. Continuations of this work carry out within the collaborative Project ASGARD (Advanced for Generation IV reactors: Reprocessing and Dissolution, contract No. FP7-CP-2011-295825).

In frame of Strategic Project "Technologies Supporting Development of Safe Nuclear Power Engineering" Domain 4 "Development of spent nuclear fuel and radioactive waste management techniques and technologies" by the National Centre for Research and Development, in INCT, works for elaborated of method synthesis of uranium carbides and nitrides doped by surrogates of MA are carried out. By using of CSGP Process, finally carbides and nitrides will gelled into irregularly agglomerates and medium sized microspheres (diameter <100 μm). The physical properties of carbide and nitride fuel make them attractive because they are conducive to high specific rod powers with relatively low fuel centre temperatures: start of life power capability is increased, power-to-melt margin is increased and fatter (more economic) pins are facilitated. Also, the very important advantages in case of transmutation is a higher actinide density to oxides. Because fuel temperatures are low, the fuel suffers little or no restructuring and the release of fission gases and volatile fission products is low [16,17].

Other developed option of fuel to the incineration of plutonium and minor actinides in thermal reactors, fast reactors, and advanced systems are Inert Matrix Fuels (IMF). The desired properties of the material(s) are a high melting point, good thermal conductivity, good compatibility with the cladding, low solubility in the coolant, good mechanical properties, and high density. In all cases, the IMF candidate is compared to standard UO_2 fuel as the reference [18,19]. A critical assessment of selection of materials as diluents for burning of plutonium fuels in nuclear reactors based on their physical and chemical properties are discussed in a recent review. This review appraises the out-of pile properties of Al_2O_3, MgO, $MgAl_2O_4$, CeO_2, Ce_2O_3, Y_2O_3, ZrO_2, $CePO_4$, $ZrSiO_4$, B4C, SiC, AlN, Mg_3N_2, Si_3N_4, CeN, YN and ZrN include, e.g. melting point, vapour pressure and thermal conductivity; the chemical behavior towards steel, liquid sodium, water, air and nitric acid in the context of suitability of diluents [20].

Figure 1. SEM analysis of uranium dioxide in the shape of irregular powders (top) and spherical particles (diameter below 100 µm) obtained by CSGP.

In the task of the Strategic Project there was elaborated method of synthesis of zirconium dioxide doped by surrogates of MA (finally plutonium and MA). This method will be used to study of nuclear transmutation of long lived actinides to short-lived nuclides.

2.4. Radioactive Waste Treatment

All fuel cycles generate some types of long-lived radioactive wastes; thus the ultimate need for disposal of those wastes. The disposal of SNF and HLW has been a major technical and institutional challenge. The modern concept of safe management of high-level liquid radioactive wastes (HLRW) is strongly connected with reprocessing of nuclear spent fuel. It provides for the necessity in fractional separation of highly active components of these wastes including ^{235}U and ^{239}Pu, minor actinides (^{241}Am, ^{244}Cm, ^{237}Np), and fission products (^{137}Cs and ^{90}Sr) followed by their immobilization or transmutation into short-lived or stable isotopes prior to disposal [21]. Selective removal of radionuclides from the bottoms residue of evaporation equipment used in NPPs has an enormous advantage over the conventional methods currently being used to condition liquid radwastes (cementing, bituminization). The advantage is primarily due to the decrease in the volume of the conditioned wastes put into solid radwaste repositories.

The nuclear industry produces large volumes of radioactive solution waste, which requires treatment prior to final disposal or storage. A highly efficient treatment concept is the removal of harmful radionuclides from the bulk waste solution. Application of such a technique will result in considerable reductions in the volumes of waste that require solidification prior to final disposal, as well as in radioactive discharges from storage containers into the environment. Number of methods are used to treat aqueous radioactive wastes, including chemical precipitation, evaporation and ion exchange, as well as less developed solvent extraction, biotechnological processes and membrane methods.

Before final storage of HLW in geological repositories, wastes have to be immobilized and encapsulated. The possible water penetration and leaching of radionuclides from stored radioactive waste makes this problem very important issue. The standard material used to encapsulate fission products and minor actinides is the vitreous matrix. The main advantages of this type of containment matrix are that it is very leach-resistant (dissolution in water) and radiation-resistant material. It is relatively easy to produce on a technical level, and it can accommodate a wide range of radioelements in compact packaging.

The most significant disadvantage of glasses for such use is their high processing temperature of ~2000°C. Sintering of sol-gel-derived glasses can be accomplished at much lower temperatures, and sol-gel techniques have been successfully used for preparation of porous glass hosts for nuclear wastes [22,23]. Appropriately prepared, sintered ceramic bodies can have higher stabilities and be more resistant to leaching than are many melt-processed glasses [24]. In one frame of Strategic Project, researches in INCT are focused on elaborating of "cold" vitrification process and immobilization HLW in ceramic matrixes. Mentioned CSGP process [25] has been adapted to synthesize silica glasses capable of incorporating significant concentrations of high-level nuclear wastes. An example flow-chart for the preparation silica gels is shown in Figure 2.

Gels in the form of powders or monoliths were prepared by hydrolysis and subsequent polycondensation of tetraethoxide/me nitrate solutions containing ascorbic acid (ASC) as a catalyst, instead of the HCl or NH_4OH that are routinely used for catalysis in glass synthesis.

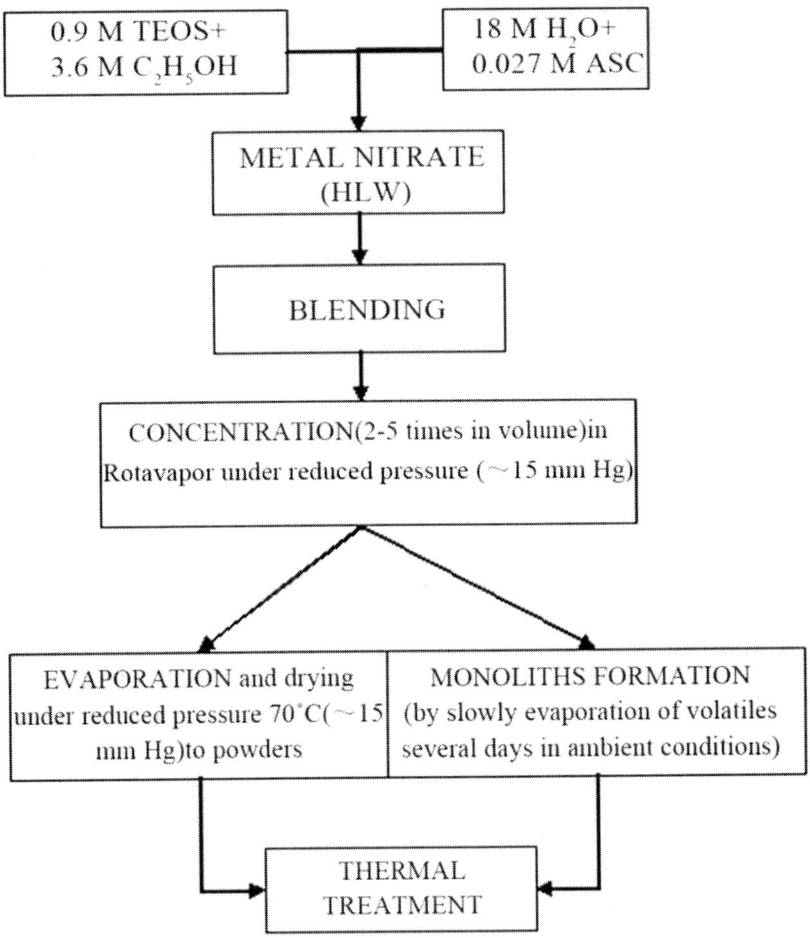

Figure 2. Flow-chart of CSGP for preparation of silica glass.

The second task of the project is immobilization of radionuclides SYNROC materials [26]. The INCT sol-gel method of synthesis titanates are used to synthesis titanate matrixes for encapsulation nuclear wastes [27].

2.5. Radioand Radiation Chemistry in INCT Poland

A chemistry program is essential for the safe operation of a nuclear power plant. Moreover in Polish situation, when Polish Government has lunched: "Program of development of nuclear energy in Poland" R & D background has a key role. The only R & D Institute in Poland which covers most of the fields of nuclear chemistry, radiochemistry, nuclear chemical engineering and radiation chemistry is the Institute of Nuclear Chemistry and Technology. The Institute runs the PhD studies in this field. Some researches of the Institute were presented earlier in the thematic chapters. Another R & D unit the National Centre for Nuclear Research (NCBJ), Świerk was established in September 2011. NCBJ is strongly involved in developing nuclear reactor engineering and promoting practical applications of nuclear physics methods. Concerning the radiochemistry major market products manufactured in the NCBJ include radiopharmaceuticals and other radioisotopes. Other activities regarding radioand radiation chemistry are ongoing at some universities. Regarding the radiochemistry, the most academic centers are working in the radioecology field Department of Radiochemistry and Colloid Chemistry, Faculty of Chemistry, Maria Curie-Skłodowska University. Lublin after Chernobyl accident focused its activities on defining a radiological state of environment — determination of the contamination level of soil, river and lake sediments, as well as ground level air with the anthropogenic isotopes such as: ^{137}Cs, ^{90}Sr and Plutonium. General research interest of the Analytics and Environmental Radiochemistry Chair at Faculty of Chemistry, University of Gdańsk, is focused on analytical chemistry and radiochemistry [28,29]; application of activity disequilibrium between $^{210}Po/^{210}Pb$, $^{234}U/^{238}U$, $^{238}Pu/^{239+240}Pu$ and $^{241}Pu/^{239+240}Pu$ as well as isotopic ratio $^{240}Pu/^{239}Pu$ for research on polonium, uranium and plutonium sources in the natural environment; impact of the Chernobyl accident on radioactive pollution; radiological risk of radionuclides intake with air, water and food consumption as well as cigarette smoking by consumers. The group working in the Department of Chemistry of the University of Warsaw has been engaged for years in the study of isotope effects on various physicochemical properties of chemical substances [30]. There are three research groups active in the field of radiation chemistry in Poland. One exists at the INCT and uses as a main experimental method nanosecond pulse radiolysis with UV/VIS detection based on the electron linear accelerator 10 MeV. The second is located at the Institute of Applied Radiation Chemistry (IARC), Technical University of Łodź equipped in PR1—pulse radiolysis system using electron beam pulses of duration variable from 2.5 ns to 4.5 μs, dose per pulse from 2 Gy to 1 kGy, wavelength spectroscopic range 250 - 2000 nm at room temperature, recorded data time range from 500 ns to 2 s FS (full scale) with 1 ns resolution time. IARC is working on some aspects of radiochemistry as well. The third radiation chemistry group exists in the Chemistry Department of the University of Podlasie, Siedlce. Some aspects of radiochemistry and radiation chemistry developments are discussed by Narbutt [13] and Chmielewski [31].

INCT in collaboration with other institutions was responsible for chemistry program preparation, personnel training and pilot studies for NPP Żarnowiec. These are well known facts that this kind of knowledge is not being delivered in the frame of the most well elaborated contract. Technology provider will not guarantee full responsibility for the operations which has to be updated to the state of art available in the next 5 - 10 years. Moreover foreign suppliers will not able to provide an adequate training of a personnel to be employed in the given country conditions.

3. CONCLUSION

This paper has been prepared to underline the role of chemistry and its contribution to the nuclear energy power development. It shows only few aspects which connect chemistry with nuclear engineering, but it is just "the tip of the iceberg". Mostly, it is referring to the INCT and author's works. However, its content illustrate well the role of chemistry in the present safe and effective operation of all stages of nuclear power related technology. Hopefully, it will pay attention on this issue and stimulate researches in this field.

ACKNOWLEDGEMENTS

This paper has been prepared in frame of Strategic Project "Technologies Supporting Development of Safe Nuclear Power Engineering" Domain 4 "Development of spent nuclear fuel and radioactive waste management techniques and technologies" by the National Centre for Research and Development (SP/J/4/143 321/11).

REFERENCES

1. International Atomic Energy Agency, "Chemistry Programme for Water Cooled Nuclear Power Plants," Specific Safety Guide No. SSG-13, 2011.

2. J. Hurwic, "Maria Skłodowska-Curie and Radioactivity," Galant Edition, Warsaw, 2011.

3. H. Polkowska-Motrenko and L. Fuks, "Proficiency Testing Schemes on Determination Of Radioactivity in Food and Environmental Samples Organized by the NAEA," Nukleonika, Vol. 55, No. 2, 2010, pp. 149-154.

4. F. Nordmann, "PWR and BWR Chemistry Optimatization," Nuclear Engineering International, Vol. 56, No. 689, 2011, pp. 24-29.

5. T. V. Epimakhov, L. N. Moskvin, A. A. Efimov and O. Y. Pykhteev, "Positive Effect of Adding Aluminum Salts to Primary Coolants of Nuclear Power Installations," Radiochemistry, Vol. 52, No. 6, 2010, pp. 581- 584.

6. J. Narbutt, "Inorganic Ion Exchangers as Selective Adsorbents and Potential Primary Barriers for Radionuclides," In: The Environmental Challenges of Nuclear Disarment, Kluver Academic Publishers, Netherlands, 2000, pp. 237-243.

7. D. Olander, "Nuclear Fuels—Present and Future," Journal of Nuclear Materials, Vol. 389, No. 1, 2009, pp. 1-22.

8. Nuclear Energy Agency, "National Programmes in Chemical Partitioning. A Status Report," NEA No. 5425, Nuclear Energy Agency, Organization for Economic CoOperation and Development, 2010.

9. J. N. Mathur, M. S. Murali and K. L. Nash, "Actinide Partitioning—A Review," Solvent Extraction and Ion Exchange, Vol. 19, No. 3, 2001, pp. 357-390.

10. C. Madic, F. Testard, M. J. Hudson, et al., "PARTNEW— New Solvent Extraction Processes for Minor Actinides," Final Report, CEA-R-6066, 2004.

11. Z. Kolarik, U. Müllich and F. Gasner, "Selective Extraction of Am(III) over Eu(III) by 2,6-Ditriazolyland 2,6- Ditriazinylpyridines," Solvent Extraction and Ion Exchange, Vol. 17, No. 1, 1999, p. 23.

12. J. Narbutt, "Hydrometallurgic Separation of Minor Actinides from High Active Nuclear Waste for Their Transmutation—Collaborative Project ACSEPT, 7. Framework Programme, EU, Euratom," V Polish Conference on Radiochemistry and Nuclear Chemistry, Kraków, 24-27 May 2009, p. 2.

13. J. Narbutt, "Trends in Radiochemistry at the Beginning of the 21st Century," Nukleonika, Vol. 50, Suppl. 3, 2005, pp. S77-S81.

14. Deptuła, M. Brykała, W. Łada, D. Wawszczak, T. Olczak and A. G. Chmielewski, "Method for Preparing of Uranium Dioxide in the Form of Spherical and Irregular Grains," Polish Patent Pending No. P-389385 (27-

10-2009), European Patent Application No. 10188438.5—1218 (2010), Russian Federation Patent Application 2010136670 (2010), Belarus Patent Application 20101305 (2010), Ukraine Patent Application 201010756 (6-09-2010).

15. Deptuła, M. Brykała, W. Łada, D. Wawszczak, T. Olczak, G. Modolo, H. Daniels and A. G. Chmielewski, "Synthesis of Uranium and Thorium Dioxides by Complex Sol-Gel Processes (CSGP)," Proceedings of the 3rd International Conference on Uranium, 40th Annual Hydrometallurgy Meeting, Vol. II, Saskatoon, 2010, pp. 145-154.

16. K. Minato, M. Akabori, M. Takano, Y. Arai, K. Nakajima, A. Itoh and T. Ogawa, "Fabrication of Nitride Fuels for Transmutation of Minor Actinides," Journal of Nuclear Materials, Vol. 320, No. 1-2, 2003, pp. 18-24.

17. International Atomic Energy Agency, "High Temperature Gas Cooled Reactor Fuels and Materials," International Atomic Energy Agency, Vienna, 2010.

18. Degueldre, et al., "Plutonium Incineration in LWRS by a Once through Cycle with a Rock-Like Fuel," Materials Research Society Symposium Proceedings, Vol. 412, 1996, p. 15.

19. H. Kleykamp, "Selection of Materials as Diluents for Burning of Plutonium Fuels in Nuclear Reactors," Journal of Nuclear Materials, Vol. 275, No. 1, 1999, pp. 1-11.

20. International Atomic Energy Agency, "Viability of Inert Matrix Fuel in Reducing Plutonium Amounts in Reactors," International Atomic Energy Agency, Vienna, 2006.

21. T. A. Maryutina, M. N. Litvina, D. A. Malikov, et al., "Multistage Extraction Separation of Am(III) and Cm(III) in Planet Centrifuges," Radiochemistry, Vol. 46, No. 6, 2004, pp. 596-602.

22. T. Woignier, J. Reynes, J. Phalippou and J. L. Dussossoy, "Nuclear Waste Storage in Gel-Derived Materials," Journal of Sol-Gel Science and Technology, Vol. 19, No. 1-3, 2000, pp. 833-837.

23. T. Woignier, J. Reynes, J. Phalippou and J. L. Dussossoy, "Sintered Silica Aerogel: A Host Matrix for Long Life Nuclear Wastes," Journal of Non-Crystalline Solids, Vol. 225, 1998, pp. 353-357.

24. R. Clarke, "Ceramic Materials for the Immobilization of Nuclear Waste," Annual Review of Materials Research, Vol. 13, 1983, pp. 191-218.

25. Deptula, W. Lada, T. Olczak, M. T. Lanagan, S. E. Dorris, K. C. Goretta and R. B. Poeppel, "Method for Preparing High-Temperature Superconductors," Polish Patent No. 172618, 1997.

26. E. Ringwood, "Immobilization of Radioactive Wastes in SYNROC," American Scientist, Vol. 70, No. 2, 1982, pp. 201-207.

27. Deptula, K. C. Goretta, T. Olczak, W. Lada, A. G. Chmielewski, U. Jakubaszek, B. Sartowska, C. Alvani, S. Casadio and V. Contini, "Preparation of Titanium Oxide and Metal Titanates as Powders, Thin Films, and Microspheres by Novel Inorganic Sol-Gel Process," Materials Research Society Symposium Proceedings, Vol. 900E, 2005.

28. Skwarzec, K. Kabat and A. Astel, "Seasonal and Spatial Variability of 210Po, 238U and 239 + 240Pu Levels in the River Catchment Area Assessed by Application of Neural-Network Based Classification," Journal of Environmental Radioactivity, Vol. 100, No. 2, 2009, pp. 167- 175.

29. Strumińska-Parulska and B. Skwarzec, "Plutonium Isotopes ^{238} Pu, $^{239+240}$ Pu, ^{241}Pu and ^{240}Pu/239Pu Atomic Ratios in Southern Baltic Sea Ecosystem," Oceanologia, Vol. 52, No. 3, 2010, pp. 499-512.

30. Makowska, A. Siporska and J. Szydłowski, "Isotope Effects on Miscibility of 1-Alkyl-3-Methylimidazolium Bis(Trifluoromethyl)Sulfonyl Imides with Aromatic Hydrocarbons," Fluid Phase Equilibria, Vol. 282, No. 2, 2009, pp. 108-112.

31. G. Chmielewski, "Chemistry for the Nuclear Energy of the Future," Nukleonika, Vol. 56, No. 3, 2011, pp. 241- 249.

Chapter 9

The Fukushima Nuclear Accident: Insights on the Safety Aspects

Zieli Dutra Thomé[1], Rogério dos Santos Gomes[2], Fernando Carvalho da Silva[3], Sergio de Oliveira Vellozo[1]

[1]Department of Nuclear Engineering, Military Institute of Engineering, Rio de Janeiro, Brazil

[2]Directorate of Radiation Protection and Nuclear Safety, Brazilian National Commission for Nuclear Energy, Rio de Janeiro, Brazil

ABSTRACT

The Fukushima nuclear accident has generated doubts and questions which need to be properly understood and addressed. This scientific attitude became necessary to allow the use of the nuclear technology for electricity generation around the world. The nuclear stakeholders are working to obtain these technical answers for the Fukushima questions. We believe that, such challenges will be, certainly, implemented in the next reactor generation, following the technological evolution. The purpose of this work is to perform a critical analysis of the Fukushima nuclear accident, focusing at the common cause failures produced by tsunami, as well as an analysis of the main redundant systems. This work also assesses the mitigative procedures and the subsequent consequences of such actions, which gave results below expectations to avoid the progression of the accident, discussing the concept of sharing of structures, systems and components at multi-unit nuclear power plants, and its eventual inappropriate use in safety-related devices which can compromise the nuclear safety, as well as its consequent impact on the Fukushima accident scenario. The lessons from Fukushima must be better learned, aiming the development of new procedures and new safety systems. Thus, the nuclear technology could reach a higher evolution level in its safety requirements. This knowledge will establish a conceptual milestone in the safety system design, becoming necessary the review of the current acceptance criteria of safety-related systems.

KEYWORDS

Fukushima Nuclear Accident, Nuclear Safety, Safety Culture

1. INTRODUCTION

In the past, TMI and Chernobyl nuclear reactor accidents were a source of operational experiences in accident conditions, pointing out new directions for all stakeholders, bringing improvements to existing power plants. The recent accident at the Fukushima-Daiichi Nuclear Power Plant, as expected, has generated reflections about the safety of nuclear power plants operation. Thus an improvement of safety systems and operating procedures becomes necessary, representing a technological challenge for the existing reactors and for next generation of nuclear power plants as well.

The Fukushima nuclear accident has produced doubts and questions which need to be properly understood and addressed, in order to enable the continued use of nuclear power technology. Such lessons, which will come from Fukushima, will establish a conceptual milestone in the safety system design, becoming urgent the review and revaluation of the adequacy of the current acceptance criteria of safety-related systems at multi-unit stations, mainly those criteria concerning to sharing of structures, systems and components (SSC).

The accident was characterized by a progressive damage due to the loss of core cooling capability, which caused the core melting. The accident progression was characterized successively by: a) Loss of off-site power supply due to the earthquake; b) Loss of on-site power supply caused by flooding due to uncommon tsunami, producing a station blackout (SBO); c) Loss of functionality of residual heat removal systems; d) Heating reactor core and its subsequent melting; e) Generation of hydrogen due to the cladding oxidation at high temperatures and posterior hydrogen detonation, damaging the reactor buildings [1] .

Several mitigation actions produced unsatisfactory results, which led to the drastic worsening of the accident, due to the existence of sharing of structures, systems and components, present in the designs of the Fukushima reactors. It is worth mentioning that this is a rather common practice in several nuclear power stations.

This paper is organized as follow: Section 2 presents the accident description, Section 3 discusses the safety culture conception, in Section 4, we discuss the design basis for flooding, Section 5 performs an approach to improve the reliability of the safety systems, in Section 6, we point out that plant-specific training is an essential safety requirement , in Sections 7 and 8, we also point out the challenge of the venting system design, Sections 9 and 10 present the concept of structures, systems and components sharing at Nuclear Power Plants (NPP) and its impact in the Fukushima accident, Section 11 presents the hydrogen explosion problem and in Section 12 we summarize some lessons learned from Fukushima accident.

2. A BRIEF DESCRIPTION ABOUT THE FUKUSHIMA NUCLEAR ACCIDENT

On March 11, 2011 a severe earthquake occurred off the northeastern coast of Japan with the epicenter about 180 km away from the Fukushima Daiichi NPP, generating a tsunami which caused devastating damage over the whole nuclear site, composed by six BWR nuclear reactors. The electric output of these nuclear reactors are of 460 MW for Unit I, 784 MW for Units 2, 3, 4 and 5, and 1100 MW for Unit 6. The reactors 1, 2 and 3 were in operation at rated power output before the event, while the reactors 4, 5 and 6 had been in an outage.

Looking for the set of effects coming from the earthquake impact, one can conclude that the automatism of the power plants have worked quite well, scramming all operating NPPs promptly. It should be noted that was mentioned that, however, there was the occurrence possibility of a small-scale loss of coolant accident at the Unit 1 [2] . This possibility is being examined further by the Japanese Government. Considering that the external electric power sources have been lost, the emergency diesel generators automatically started, supplying power to the residual heat removal, before the tsunami arrival.

The tsunami impact, however, caused deep operational and structural damages, becoming inoperative the diesel generators and batteries. Without AC or DC power to the energy supply to the operation of decay heat removal systems, the fuel of nuclear reactors 1, 2 and 3 suffered total and partial melting respectively.

A large amount of hydrogen was released, in consequence of the reaction between zirconium and high temperature steam water, producing strong explosions, damaging the buildings of the Units 1, 3 and 4. It should be noted that the Unit 4 was shutdown and had been in a refueling outage. All of the nuclear fuel had been removed from the reactor and placed in the spent fuel pool. Because of hydrogen explosion at Unit 4, it was initially supposed that the spent fuel was uncovered, and therefore, producing hydrogen by oxidation. Despite this previous hypothesis, subsequent assessments conclude that the spent fuel stored at the pool was not damaged by the accident.

In November 2011, the Xenon-135 detection at the Unit 2 opened doubts about the evidence of nuclear recriticality, however, was concluded that the spontaneous fission process of actinides, from the damaged fuel, was the unique responsible for the xenon production [3] .

3. NUCLEAR SAFETY CULTURE

The International Nuclear Safety Advisory Group (INSAG) in its report about the Chernobyl accident assessment coined the term safety culture to refer to the safety regime that should prevail at a nuclear plant operation, which was introduced globally to explain how the lack of knowledge and understanding of risk and safety by the employees and organization contributed to the Chernobyl disaster [4].

Later, INSAG, under the auspices of the International Atomic Energy Agency (IAEA), enlarged the safety culture concept, considering that the development of a safety culture must be embedded in the national legislative and regulatory framework, establishing the proper chain of responsibility and authority for the required level of safety. In both operating and regulatory regimes, safety culture must be instilled in organizations through proper attitudes and practices of management [5].

A review of nuclear incidents indicates that safety culture problems affect both highly developed and developing countries [6] . Safety culture issues can arise at all stages of organizational life, and even in organizations previously recognized for their safety performance. Currently the majority of effort to improve safety culture has focused on nuclear power plants.

Looking the Fukushima accident, it seems evidenced the need of improvements in the Japanese safety culture, taking into account the lessons learned from Three Mile Island and Chernobyl accident, as well as the tools development aiming to monitor its adequate implementation. So, must be highly recommended to perform international audits programs (peer reviews) covering both operational and regulatory regimes, aiming to strengthen and enhance the effectiveness of the safety culture taking into account the IAEA safety standards and the good international practices.

4. BARRIER PROTECTION AGAINST TSUNAMI

Hydrological phenomena, such as flooding due to tsunamis, can cause several hazards that could affect the safety of nuclear power plants, leading to the risk of common cause failure for essentials safety systems, such as the emergency power supply systems or the electric switchyard, with the associated possibility of loss of off-site power, the decay heat removal system and other vital systems [7] [8] .

IAEA presents that a conservative analysis of the tsunamis effects should be made for inclusion in the Nuclear Power Plants (NPP) design basis, taking into account the estimation of probable maximum tsunami (PMT), aiming to protect NPPs against all potential effects of external events, determined by using historical data and geological, tectonic and seismic investigations [7] .

The licensing design basis for flooding generated by tsunamis of Fukushima Daiichi was initially estimated based on the effects at Japan of the tsunami generated by the magnitude 9.5 Chile earthquake in 1960. Thus, it was defined one PMT producing waves with up to 3.1 meters over mean sea level [9] .

Later, new methodology to estimate tsunamis were developed by the Japan Society of Civil Engineers, in 2002, based on observations from Shioyazaky-oki earthquake (magnitude 7.9) in 1938, which resulted in a maximum water level of 5.7 meters [9] [10] . This value was not reviewed or validated by the Japanese Nuclear Regulatory Body (NISA). This assessment was undertaken by plant's operator (TEPCO) voluntarily without any instruction from NISA, and therefore, not officially recognized in the licensing documents [9] .

This estimative corresponds to the tsunami height at the shoreline point in the entrance to intake structures level. The run up i.e. the water height reached at the maximum inundation point was not indicated in any presentation from TEPCO. It seems also that the calculation of the run up have not considered the specific and detailed arrangements of plant layout [9] .

In the beginning of this century, geological evidences allowed to know the occurrence of the Jogan earthquake in 869 AD, with an estimated magnitude of 8.6. This earthquake caused a giant tsunami in the region of Sendai, in the province of Fukushima. According to estimates arising from the geological sediment analysis, the tsunami waves of 869 AD, similarly to the tsunami of

2011, penetrating up to 4 km from the coast, causing a great flood [11] , however, the design-basis flood were not updated to take into account this new historical data.

It must be emphasized that was presents that, since 2006, NISA and TEPCO as well were aware of the possibility of a SBO at the Fukushima site if the nuclear site was reached by a tsunami and that NISA knew that TEPCO had not prepared any measures to lessen or eliminate the risk, and failed to provide specific instructions to remedy the situation [2] .

It is practically consensus that the Japanese nuclear regulatory body and the plant operator did not follow international best practices and standards concerning defenses against extreme external events. Thus, the resistance of Fukushima-Daiichi NPP to tsunamis was underestimated [2] [11] .

It must noted that during the assessment for life extension of Unit I, NISA has not imposed, as requirement, the update of flood studies related to the protection of tsunamis, with the inclusion of the Jogan earthquake at its design basis.

During a meeting in London, on July 2011, the Vice-Chairman of the Japan Atomic Energy Commission noted that in recent years several warnings had been issued by the Japanese academic community about the vulnerability of nuclear power plants for the earthquakes and tsunamis occurrences. He also noted that, in 2009, some members of NISA have questioned the non-inclusion of the Jogan earthquake in studies to the design basis updating of Japanese nuclear power plants [12] .

5. DIVERSITY AND REDUNDANT SAFETY SYSTEMS

Among several definitions, redundancy is a common approach to improve the reliability and availability of a system, thus enabling that safety systems adequately work, even if an individual item fail to perform on demand. It should be noted that the use of redundancy increases the cost and complexity of a system design, however, it is possible to postulate an event that could produce a common cause failure (CCF) inducing failures in two or more channels of a redundant system, which can lead to its inability to function as designed.

The proper application of the concept of diversity is the way to protect redundant systems against CCF. The diversity includes the differences between the system components, considering design, manufacturers, installation, software, operation and maintenance procedures and the differences in their environment and location, in order to prevent the lost of their redundant functions [13] .

The off-site power was lost when the earthquake occurred and the emergency diesel generators started up throughout the power station, as expected, however, the on-site power supply redundant systems (emergency diesel generators and batteries) were lost due to flooding caused by tsunami. Due to the lost of DC power, the instrumentation became unavailable to determine the main operational parameters of all reactors or to remotely actuate valves powered by DC power, except Unit 3 [1] .

Concerning the location diversity for redundancies, is consensual that the installation of diesel generators and batteries at different levels at the Fukushima nuclear site, could reduce the vulnerability of this system to flooding, ensuring the survival of power generation system and, consequently, making feasible the operation of residual heat removal system.

It must be noted that was priority the power supply recovery, connecting batteries in series to the terminals of the control panels, in order to open valves and restore the measuring instruments, but there were no batteries stored at the Fukushima Daiichi site. However, this was improvised through the removal of batteries of the private cars of employees and TEPCO's service vehicles [14] .

Concerning the design diversity for redundancies, Fukushima Daiichi NPP had nine water cooled and three air cooled diesel generators. At this point, it is important to note that the sea water pumps, used to provide cooling to the diesel generators, were installed four meters above sea level, becoming, therefore, inoperative due to the flooding.

It should be noted that some air cooled diesel generators (DG) in the Fukushima Daiichi NPP were not damaged by the tsunami, because they have been installed on an upper floor, however, their metal-clad switchgears were flooded because was installed on the lower floor, not permitting the use of these DGs.

As one of the key lessons was that, in case of CCF, equipment reserves and essential resources, such as batteries, small generators and air compressors, should be available, in real time, to safety systems operation. If these components were readily available, the Fukushima accident could have been effectively mitigated. So, the possibility for replacing such equipments should be incorporated to the redundant system.

It seems to indicate that should be recommended the immediate availability of DC power supply, through independent batteries bank placed near the control room, allowing the opening and closing of valves of essential safety systems, such as venting and safety relief valves. An independent batteries bank should power supply to others equipments of instrumentation and control.

Additionally, should be important the installation of small diesel generators in order to recharge this battery bank, extending the battery lifetime. It must be noted that the absence of electrical power supply triggered a series of human failures (including decision-making) and equipments failures that have proven unreliable when activated. It must emphasized that the equipments qualification in aggressive environments is an essential prerequisite for the nuclear reactors design, concerning of the instrumentation and control, which would provide greater support for decision-making during the management of severe accidents.

The concept of redundant systems of nuclear power plants should be revised and expanded to incorporate the new lessons learned from this accident. New safety system technologies are under study. It is important that an international effort, by different manufacturers, be encouraged in order to use passive and active redundancies to the next generations of reactors under licensing [15] .

6. TRAINING

Nuclear safety is also closely linked with the technical and operational staff capacity and its ability to convert these experiences into operational procedures. For a proper severe accidents management, it is necessary that the personnel involved in performing the various actions during the accident conditions be adequately prepared to take effective on-site actions to prevent or to mitigate the consequences of such accident.

It is necessary that these personnel be acquainted with expected nuclear plant behavior beyond of design basis conditions, and their consequences as well. In this way, the computational simulator is an important tool for the operational training. One must be emphasized, however, that the time decision-making is closely related to familiarity with the operational characteristic of each system.

On the other hand, the management decisions are considered critical in the sense that the implementation of a decision can also include adverse effects. In such case, the potential negative consequences have to be properly assessed before an action is performed.

The Investigation Committee on the Accident at Fukushima (hereinafter referred to as the “Investigation Committee”) quoted that the shift staff had not realized that all isolation condenser system (IC) valves were designed to be closed by the failsafe function when all power supplies were lost. According this reference, the shift team could not identify the operating status of the IC immediately after the tsunami [14] .

Before the earthquake, no one in the shift staff, in charge of Unit 1, had knowledge about the operation characteristics of the IC, ignoring that a visual inspection of steam blow-out (and operating noise) could point out the operational status of IC [14] .

As mentioned before, it becomes clear that the weak familiarity of operators with rapport the main characteristics of safety systems in emergency scenarios, inducing evaluation errors and mistaken making-decision. For example, it was inhibited the water injection in the reactor core (Unit 1) during the early moments after the tsunami, causing its melting in an unexpected short time due the mistakes about the operational status of IC.

It should be noted, that the plant-specific training is an essential requirement, since it allows a deeper knowledge of main safety subsystems behavior. This accident showed the need for a continuous and effective plant-specific training in preventive and mitigative actions on simulators, which should take into account the identification and assessment of severe accident scenarios and its progression as well. One must be emphasized, therefore, that the time decision-making is closely related to familiarity with the characteristic of each system.

7. VENT SYSTEM FAILURES

The free volume of the BWR Mark I containment is quite small relative to the other containment types. Therefore, during an accident, the containment can be overpressurized in a relatively short period of time unless steps are taken to mitigate the pressure rise [16] .

Under severe accident conditions, such as core meltdown, the pressure of the containment must be released to the atmosphere to maintain the containment integrity. However, although the venting decreases the pressure in the containment, its implementation releases radioactive material into the environment.

One of the most comprehensive studies of containment venting was performed on the Peach Bottom Atomic Power Station, a BWR with a Mark I containment. A main conclusion from that study was that the containment venting had limited potential for reducing the risk associated with the dominant severe accident sequences [17] .

In 1989, the United States Nuclear Regulatory Commission (USNRC) recommended, to reduce the vulnerability of BWR Mark I containments in case of severe accidents, by the installation of a hardened wetwell vent, improving the venting capability [18] . In general, USNRC proposed a vent modification through the installation of a direct vent path from the torus to the main stack (Direct Torus Venting System—DTVS) by-passing the SGTS.

Following the USNRC recommendation basic design, each individual plant owner designed and installed the hardened vent to meet its best plant-specific design criteria. Operators in Japan, including TEPCO, did the same [19] , but there is insufficient information to know the venting system installed at Fukushima.

In the basic design, the DTVS must be isolated from the SGTS by installation of air-operated valves and a rupture disk. It should be noted that the valves of this system are normally closed and in case of loss of electrical and pneumatic power they are automatically closed. Additional filters were not considered by USNRC, because was concluded that the scrubbing at suppression pool, during the venting, was effective on reducing the source term released to the environment.

As the SBO consequence, the venting line was manually configured, through of the manual opening of vent valves, into reactor building, which required the installation of a DC power supplier and portable air compressor. It was mentioned that this manual opening is complicated due to the high radiation level that would be present following the reactor vessel failure, not allowing the access to these valves [16] . It should be noted that the venting implementation at Fukushima was interrupted sometimes by the high levels of radiation inside the reactor building [14] .

As observed in Fukushima accident, the venting action did not work properly: i) In Unit 1, the venting operation produced a hydrogen explosion in the containment building; ii) In Unit 2, the recovery staff was not able to configure the venting line, thus the core depressurization, through safety relief valves (SRV), was performed, damaging the primary containment by over-pressure. The most probable over-pressure failure location in Mark I containments is the upper half of the pressure suppression pool torus. Let us comment that, on March, 15 was heard a strong sound coming from Unit 2, and its suppression pressure equalizing with atmospheric pressure, so, indicating a suppression pool rupture as expected [20] ; iii) In Unit 3, the venting operation produced hydrogen explosions in the Units 3 and 4 due to backflow into the Unit 4 through the SGTS piping, since the exhaust pipe of unit 3 and 4 sharing the same main stack [21] .

After the venting beginning at units 1 and 3, a strong hydrogen explosion occurred on the roof of these reactor buildings. The hydrogen transport and its accumulation in the roof are not clear, as well. One possibility was that had a leakage of hydrogen through drywell head, due to the containment pressure, exceeding its design basis. However, another possibility would be the gas migration into the reactor building through the SGTS, during the depressurization. It should be noted that with SBO occurrence, the valves of SGTS should be automatically closed.

The Investigation Committee related that the operation performed, by the recovery staff to the venting configuration, was through of the manual opening of a motor-operated valve near the stack and the opening of two air-operated valves (a larger and a smaller one) in parallel pipes [14] . As recommended by USNRC, the DTVS design provides a direct vent path from the torus to the stack bypassing the SGTS through parallel pipes [18] .

According to before mentioned, we are supposing that the simultaneous opening of the parallel air-operated valves allowed the venting through two paths, via DTVS and via SGTS. The ventilation pipes and filters of SGTS were not designed to withstand large internal pressures, so these pipes would likely fail after the vent valves were opened, allowing radioactive steam, fission products and hydrogen to escape into the reactor building, with the hydrogen accumulating at the roof and likely producing an explosion [16] .

Investigating the hydrogen explosion occurred in Unit4, TEPCO concludes that the containment venting from unit 3 flowed into the Unit 4 through the SGTS pipes, since the exhaust pipes of unit 4 joins the Unit 3 exhaust pipes at the main stack. It must be emphasized, however, that the Unit 4 was in outage at the moment of earthquake [21] .

It is unclear why the isolation valves in the SGTS at the Unit 4 was opened, allowing the gas backflow into the reactor building and the subsequent hydrogen explosion, since this valves should be automatically closed due to loss of electric power, ruling out the hydrogen migration possibility from Unit 3 to Unit 4. Probably, this fact would point to a possible failure in the design of the DTVS installed in Fukushima or a mistake on the manual alignment of the vent valves, with the opening of SGTS valves.

8. FILTERED VENTING

As before mentioned, under severe nuclear accident conditions, the SGTS do not have operational capability to perform gas filtering during the venting [16] . So, when the venting was performed through suppression pool (torus), the pool scrubbing was the only filtration mechanism. It should be noted that there are another venting path, where the gases in the containment vessel are directly released to the main stack, bypassing the beneficial effects of scrubbing through the water in the torus. Therefore, the first path is preferable due to minimization of radioactive releases [14] .

It was mention that the installation of a filtered vented containment system to an existing nuclear power plant has been suggested as one approach to mitigating the effects of a severe accident, since reduces considerably the radioactive releases to environment during venting operation [21] .

Several manufacturers had presented filtered venting systems designs as an update to NPP safety system. It must be emphasized that filtered venting should have, among their main requirements, the operation without electric power, monitoring of adequate performance parameters in the control room to provide status of system during operation, as well avoid potential hydrogen detonation system is initiated following a severe accident [21] .

Additionally, it was commented that the Japanese were interested in install, in the 90´s, filtered venting systems for their nuclear plants, but its implementation was not done [22] . OCDE shown the filtered venting systems description selected for implementation in Germany, France and Sweden [23] .

The high cost/benefit has discouraged the filtered venting system installation during the last thirty years, for almost all NPPs. However, looking for the consequences of Fukushima, we believe that the mandatory implementation of filtered venting will become, probably, a tendency by the main international nuclear regulatory bodies. The desired characteristics for the filtered venting system should be studied in depth by each regulatory body and widely discussed internationally, in order to produce a consensus in the nuclear sector.

9. STRUCTURES, SYSTEMS AND COMPONENTS SHARED AT NUCLEAR POWER PLANTS

The SSC sharing aims, mainly, to reduce the construction, operation and maintenance costs. This procedure is widely used by the nuclear industry and other technologies. The SSC sharing between units at nuclear power plants translates into a reduction of overall construction time and in the respective costs for equipment, materials, and structures [24] .

The reduction of these costs may have implications for sensitive items related to nuclear safety, involving risks which must be quantified. Regardless of the material benefits, safety cannot be, however, compromised, that is, the SSC sharing cannot impair the ability of the systems to perform their safety functions. Thus, an important objective of sharing systems is to reduce cost without disturbing the nuclear safety integrity.

Due to safety, operability, and license ability reasons, the choice of SSCs to be shared cannot be exclusively based on economic assessments. So, a probabilistic risk assessment must be performed taking into account the safety impact due to the SSCs sharing and its consequences at multi-units station, compared to results obtained when applied for a unique reactor site.

The existence of sharing increases the likelihood of scenarios that could impact a single unit independently, and creates a potential for scenarios that may involve several units at the site, as observed at Fukushima.

According to Oak Ridge National Laboratory, there are three different sharing strategies. The first one is a single SSC that supports both units simultaneously. In the second one there are independent SSCs for each unit. They are interconnected to attend both units. This sharing configuration increases the availability of SSCs [24] . It is important to note that these SSCs should be designed to support the demand when one unit is lost, avoiding the system shutdown due to overload. In the last type there are independent SSCs for each unit, but sharing of standby or spare equipment.

It should be noted that, before Fukushima, the simultaneous occurrence events assumption, at multi-units stations, was not considered. Thus, in this scenario, the availability of standby equipment and resources could be, eventually, not correctly estimated. The USNRC determined that, because of the low probability of a severe reactor accident, a suitable design basis for multi-unit nuclear power plants was the assumption that an accident occurs in only one of the units at a time, with all remaining units proceeding to an orderly shutdown and a maintained cooldown condition [25] . It must be emphasized, however, that such scenario did not happen at Fukushima accident, where several units were impacted simultaneously.

The USNRC had already limited the opportunities for sharing of onsite power systems at multi-unit power plant sites, because the sharing generally results in a reduction in the number and capacity of the onsite power sources to level below those required for the same number of units located at separate sites [25] . Thus, one must question whether the SSC sharing would weaken the functionality of the redundant installed devices. This could affect the availability and proper use of these shared devices during accident mitigative actions.

As it is well known, redundancy is a common approach to improve the reliability and availability of a system, enabling safety systems performing satisfactorily its functions, even if individual systems fail when demanded. It should be noted that the use of redundant devices increases the cost and complexity of a system design. At this point, we question if the SSC sharing could be considered as in conflict to redundancy concept.

10. IMPACT EVALUATION OF THE SSC SHARING IN THE FUKUSHIMA ACCIDENT

The Fukushima accident produced many issues that need to be answered. This knowledge needs to be incorporated in the new safety systems to be implemented in next reactor generation designs. The current safety design must also be updated to consider the new requirements that are coming from Fukushima. It should be noted that this represents a challenge for all stakeholder, including the nuclear industry and the nuclear regulatory bodies.

As explained previously, before the Fukushima, it was determined that, because of the low probability of a severe reactor accident, a suitable design basis for multi-unit nuclear power plants was the assumption that an accident would occur in only one of the units at a time. It should be noted that, in a technical analysis, events that are relatively improbable cannot be ruled out, creating the risk of not having resources for their mitigation.

During the mitigation of the Fukushima accident, it is possible to observe the influence of sharing SSCs on final outcome of the accident. Thus, in a technically complex scenario some points should be analyzed separately, such as the current SSCs sharing methodology at multi-unit stations.

In subsequent section we aim to evaluate some Fukushima SSCs shared that impacted on the ability of carrying out actions to mitigate the consequences of the nuclear accident.

10.1. Control Room Sharing

There was a main control room shared by adjacent reactors at Fukushima Daiichi NPP; one for Units 1 and 2, one for Units 3 and 4, and another for Units 5 and 6. For each control room there was only one shift supervisor with the responsibility of making decisions during the course of the accident concerning the control and operation of both units, and reporting all basic information necessary to the emergency response organization [14] .

For example, there are several multi unit US nuclear power plants where there are two complete control rooms that share the same shift supervisor. In the same way, at United Kingdom the AGR reactors sites have two reactors, sharing control rooms, supervised by the same shift supervisor. Therefore, the structural arrangement designed in Fukushima Daiichi control rooms is not uncommon.

According to the Investigation Committee, communication failures between the shift staff and the on-site Emergency Response Organization (ERO) caused a misunderstanding concerning water injection in the unit 1. The ERO assumed that there was wrong evaluation of IC status. Thus, the ERO believed that the situation in Unit 2 was more dangerous than that in Unit 1. So, efforts were focused by the staff for Unit 2, instead of Unit 1, delaying the decision concerning alternative water injection into Unit 1. As a consequence of the delay in implementing of mitigative actions in Unit 1, the reactor accident at Unit 1 rapidly progressed to total fuel melting, about six hours after the tsunami arrival [14] .

The control room sharing by different Units at Fukushima Daiichi impacted the accident mitigation actions. For example, the accident mitigation actions for Unit 2 had priority over the Unit 1, so, impairing the cooling of this reactor that produced a fast melting.

As observed, it must be emphasized that, according the USNRC's acceptance criteria [26] , the control room should be considered a structure not subject to sharing. This situation is aggravated when the units shared, have different safety systems, as existent at Fukushima Units 1 and 2.

Due to the accident progression in the Unit 1, high radiation levels were seen inside the main control room, shared between Units 1 and 2. Because of this, it was necessary to avoid certain areas at the main control room [14] . It must be noted that, the staff access and permanency in Unit 2 control room could become prohibitive due to the high radiation levels. Thus, it is evident that the accident in one unit could impair the other unit due to the control room sharing.

It is important to remember that before Fukushima accident, the assumption of simultaneous occurrence of a severe accident at multi-units was not considered. So, the control room sharing, supposedly, would not disturb the activities of each team individually to perform their operational functions. This was not, however, confirmed effectively.

Concerning previous USNRC sharing recommendation, we understand that it must be revised taking into account the Fukushima accident lessons. This subject must be deeply analyzed due to its operational consequences, concerning human factors engineering and human performance in order to evaluate and to improve the safety, the efficiency and the robustness of these work systems.

10.2. Main Stack Sharing

The Fukushima Units 3 and 4 shared the same main stack which caused undesirable interactions among these units due to physical interconnection between them. Similarly, the Units 1 and 2 shared another main stack.

With the accident progression the containment pressure at Unit 3 reached values above its structural design limits, necessitating venting to the main stack, aimed at gradual depressurization containment [27] . Although the venting reduces the containment pressure, maintaining its integrity, its implementation releases radioactive materials to the environment.

Since the venting pipes of Units 3 and 4 shared the same main stack, the existence of this interconnection (as a consequence of its sharing configuration) made possible hydrogen transport into the Unit 4. The hydrogen accumulated and caused an explosion in Unit 4, damaging its reactor building, despite the Unit being in outage at the moment of the earthquake [21] . It is unclear why the Unit 4 isolation vent valves were opened, in disagreement with its operational logic. Indeed, there was an operational mistake, due to the physical status of the valves (opened), allowing the hydrogen flowing into its reactor building mainly due to pressure difference among the shared venting pipelines.

The existence of an inappropriate main stack sharing in the reactor design allowed that a single failure in the venting valves to damage Unit 4, increasing the accident consequences at the Fukushima Daiichi site. Looking for the economic benefit obtained with the main stack sharing, when faced with the observed damages; it allow us to conclude about the inadequacy of this sharing design. Thus, individual vent pipes, including the main stack, without shared devices, seems to be the best design choice for multi-unit stations.

10.3. Emergency Staff Sharing

According to the IAEA report [28] , the emergency staff members are composed of evaluators, decision makers and implementers, each one with their specific responsibilities. Looking for Fukushima, the ERO staff was poorly sized to cover the occurrence of simultaneous events at all units. This inadequacy of technical support staff led to misunderstanding, failures and delays in decision making, mainly as a consequence of personal stress. This situation becomes more severe in the case of the same staff is requested to perform actions at plants with different safety configuration designs, as for example, different water injection systems installed in Units 1 and 2.

Considering the technological differences of the Fukushima nuclear reactors, it is necessary to have unit-spe- cific training for accident conditions, including all possible scenarios. The staff should be familiar with the expected nuclear plant behavior beyond of basis design conditions and the consequences as well. This could be done using simulators for the operational training. It is more evident the need for specific training of the emergency staff for each different technology of the reactors at multi-unit stations.

Emergency staff sharing compromised implementation of actions needed to minimize the consequences of the accident. Failures in accident management occurred in all the damaged Fukushima units. For this reason, it is suggested that there be a dedicated emergency staff for each unit at multi-unit stations, in order to avoid loss of focus in case of emergency.

10.4. Resources for Emergency Actions

In Fukushima, it was observed that there was not essential equipment of resources required for emergency actions, such as batteries, small AC generators and air compressors, which should be available promptly for mitigative actions. The non availability of these resources was based in the very low estimative of one event affecting simultaneously several units at the site.

It was believed that the interconnection among the various systems shared would present a robust redundancy, thus it would not need the availability of these additional resources. It should be emphasized that accident scenarios, even improbable, cannot be discarded at a probabilistic risk assessment, otherwise they will not have adequate conditions and the resources required for mitigation if the event occurs.

Additionally, according to IAEA report [28] , water is a resource that should be available between the elements necessary for severe accident mitigation after the loss of water injection capability, when the core damage became inevitable. So, it is necessary alternative water injection in order to flood the drywell in an attempt to preclude melt through of the reactor vessel [29] .

As a result of the rapid accident progression, the molten fuel started to damage the reactor pressure vessel at the Unit 1 five hours after the tsunami arrival [11] . Thus, possible mitigative actions to maintain the fuel in the vessel became impossible.

It should be noted that the only alternative for water injecting at Fukushima were the pumps installed on fire engines. It should be emphasized that there was only one fire engine to attend all damaged power plants of this site, since that one was destroyed by the tsunami and another fire engine remained stopped between units 5 and 6 because of the destruction of the internal roads [14] .

The existence of only three fire engines to attend six nuclear power plants at Fukushima can be understood as a resource sharing that proved to be inadequate to act in multi-unit accident scenario. It should be noted that ORNL describes the use of portable pumps provided by fire engines as a possible tool for alternative water injection into the core during mitigation actions, keeping the damaged core inside the reactor vessel and avoiding its subsequent failure and consequently the radioactivity release to the environment. It should be noted that these pumps must have adequate capacity and enough power to proper use as an alternative safety system [29] .

The non-availability of these resources, in quantity and adequate capacity, disturbed the making-decision in due time, thus, implying that the performed actions have been compromised.

11. HYDROGEN EXPLOSIONS

It should be emphasized that until the explosion at Unit 1, no one at the Fukushima Daiichi NPP, TEPCO Head Office or the Japanese Government considered the possibility of a hydrogen gas explosion occurring in the reactor building [14] . Thus, not having been considered this possibility of explosion, no possible mitigation action, related to hydrogen concentration, was done, until the explosion in Unit 1. Later, new hydrogen explosion occurred in units 3 and 4, as mentioned before. Thus, failures occurred concerning to the identification of the accident progression sequence, although already available in the literature. It should be noted, however, that mitigation strategies of hydrogen explosions were very limited.

During the progression of the severe accident, the reactor core temperature increases, due to inoperability of the residual heat removal system, resulting on large amounts of hydrogen generation, due to zircaloy cladding oxidation reaction with steam water at high temperatures.

So, it becomes evident that an important scientific challenge for the safety use of nuclear energy (for electricity generation) should be the minimization of hydrogen generation, in the case of the occurrence of severe accidents.

The hydrogen combustion can cause explosions that may damage the containment building. Since hydrogen combustion represents a hazard to the containment integrity, mitigation strategies for hydrogen becomes one of the essential parts of any accident management program. The controlled venting, in an early stage of the accident, is assumed to prevent high hydrogen concentration in the containment [30] .

In most countries, there are no strict regulatory requirements on the implementation of hydrogen mitigation strategies for existing plants. However, for the new reactors that are planned or under construction, these considerations must be included into the design [30] .

Some solutions have being considered, as the passive autocatalytic recombiners (PAR) for mitigation of hydrogen generated during a severe accident (due to zircaloy oxidation at high temperature). These devices recombine hydrogen with oxygen producing steam and heat. This exothermal reaction may lead to an overheating of the catalyst elements and consequently cause an unintended ignition of the hydrogen/air-mixture. However, research is ongoing to create PARs with reduced probability of hydrogen ignition [31] .

Another mitigation strategy is the deliberate ignition system to initiate combustion wherever and whenever flammable mixtures arise, removing the hydrogen by slow deflagration. It should be noted that some works have recommended the use of dual concept, using integrated recombiner-igniter system for the gas control in order to cover a broader spectrum of accident sequences, and to provide adequate diversity [32] [33] .

The search of alternative materials, for replacing the zirconium in the fuel cladding, is in progress. It is a way to prevent (or minimize) the hydrogen production. Although all metals react with steam water at high temperature, the stainless steel would be an option better than Zirconium alloys, regarding the resistance to oxidation. Another option being considered is the use of ceramics materials, such as silicon carbide aiming the hydrogen production minimization [34] . The material eventually chosen must have adequate neutronic properties for use in nuclear reactors.

In Fukushima accident, the implementation of an early venting was not possible because of the loss of electric energy and pneumatic supply. There were no other procedures or available equipment for mitigation strategy. The later completion of the venting, after the manual configuration of its path, produced strong hydrogen explosions in units 1, 3 and 4.

Additionally, it is important that the concentration of hydrogen in the containment must be permanently monitored. Some accident management procedures and making decisions may depend directly of the hydrogen concentration value. There were no gauges to hydrogen concentration measurement at the Fukushima NPP [14] .

As already mentioned, the Hydrogen problem must be rethought, remaining as a challenge for the nuclear manufacturers and for the scientific community.

12. LESSONS LEARNED

The SSCs sharing has shown that mitigation actions do not produce the expected results in order to prevent the evolution of the accident, being worsened due to low familiarity of the emergency staff with the safety systems. Thus, it is important that the periodic training program must include beyond basis design scenarios. So, the emergency staff must have full knowledge about the safety-related systems. It is evident that the possibilities of sharing at multi-units site should be carefully reanalyzed, since its applicability becomes more restrictive.

The main events that occurred during the Fukushima accident were considered, indicating the need of improvement of techniques and procedures to mitigate its consequences. So, the lessons learned from this accident are summarized below.

Lesson learned 1—A complex failures scenario occurred at Fukushima, including omissions of nuclear regulatory body, due to lack of compliance with IAEA safety standards and good international practices, allowing an underestimation probability of external events occurrence. Thus, it is highly recommended to perform international technical audits programs, aiming to strengthen and enhance the effectiveness of the safety culture.

Lesson learned 2—The concept of electrical interconnection between all power plants units, like existent at Fukushima, cannot be understood as a robust redundancy system to support common cause failure events, thus a most adequate use of diversity concept is the effective way to protect redundant systems minimizing the undesirable occurrence of SBO. Beyond this, the concept of redundant systems of nuclear power plants should be revised and expanded to incorporate the lessons from this accident. Indeed, this new redundant systems must be most complex, embedded in the concept of defense in depth.

Lesson learned 3—It seems to indicate that should be recommended the immediate availability of DC power supply, through independent battery banks, placed near the control room, allowing the opening and the closing of valves of essential safety systems, such as venting and safety relief valves, and also the power supply to others equipments of instrumentation and control. Additionally, should be important the additional installation of small diesel generators in order to recharge this battery bank, extending its lifetime.

Lesson learned 4—It must be noted that the absence of electrical power supply triggered a series of human failures (including decision-making) and equipments failures that have proven unreliable when activated. It must emphasized that the equipments qualification in aggressive environments is an essential prerequisite for the nuclear reactors design, concerning of the instrumentation and control, which would provide greater support for decision-making during the management of severe accidents.

Lesson learned 5—The severity of the Fukushima accident points to the necessity of an international consensus among all stakeholders, concerning the increase of nuclear safety conditions, thus it should be evident that after the accident in Fukushima, it becomes necessary to define new safety assessment criteria, including acceptance criteria for the use of SSCs sharing, taking into account the possibility of simultaneous accident occurrence at several reactors at multi-unit stations. The use of sharing for costs reduction may impair the nuclear safety. So, the sharing involving risks must be quantified. Regardless of the material benefits, safety must not be compromised. Due to nuclear safety reasons, the SSCs sharing design must not be exclusively based on economic assessments.

Lesson learned 6—At Fukushima there was a main control room for adjacent units. In this way, in the first moments of the accident, efforts were focused for unit 2 instead the Unit 1, delaying the making-decision concerning to beginning of an alternative water injection into Unit 1. Thus, this work concludes that each control room should be considered a structure not subject to sharing, avoiding the hierarchical competitive relating making-decision among power plants, during the mitigative actions.

Lesson learned 7—It was identified inadequate training for emergency situations, resulting in delay in making-decisions and deterioration of the situation due to loss of time windows to apply the mitigative procedures. So, preventive and mitigative actions training can be performed on simulators, which should take into account the identification and assessment of severe accident scenarios and its progression as well.

Lesson learned 8—It should be noted that the lack of support staff, of resources and of equipments not allowed that the mitigative actions were adequately performed, take into account a simultaneous accident scenario in several Units of Fukushima-Daiichi NPP. As observed in this study, there was a high degree of improvised actions to prevent the progression of the accident, since the AMP documentation did not contained procedures to be followed in accident conditions due to external events.

Lesson learned 9—It must be emphasized that venting actions do not work properly at Fukushima, thus its design must be rethought. However, looking the accident consequences we believe that the mandatory implementation of filtered venting will become a tendency by the main international nuclear regulatory bodies.

Lesson learned 10—The Hydrogen problem must be also rethought, remaining as a challenge for the nuclear manufacturers and for the scientific community. It is in progress the search of alternative materials for replacing the zirconium in the fuel cladding as a way to minimize the hydrogen production. However, some solutions already available, as the autocatalytic recombiners and igniter systems, could be used to hydrogen concentration control in primary containment.

Lesson learned 11—Due to the main stack sharing, the hydrogen migration from unit 3 to 4 caused an explosion, destroying the Unit 4 reactor building, despite being in outage. Regarding this accident, the sharing of venting pipelines, including the main stack, should be avoided at multi-unit stations design.

Lesson learned 12—It should be emphasized that independently of the safety system improvements is essential the evolution on the personnel knowledge to actuate during a severe accident. These efforts should be focused the development of training programs, with massive use of simulators as the important tool for the operational training, in severe accident scenarios.

13. CONCLUSIONS

This work performed a critical analysis of the Fukushima nuclear accident, focusing at the common cause failures produced by tsunami, as well as an analysis of the main redundant systems. It also assessed the mitigative procedures implemented during the accident and the subsequent consequences of such actions. Our analysis shows that the accident management gave results below expectations, not avoiding the progression of the accident.

We did show that the inappropriate use of concept of SSC sharing in safety-related devices at multi-unit nuclear power plants had a negative effect on the nuclear safety, contributing to the Fukushima accident scenario.

The NPPs of the world are waiting for the technical answers for the Fukushima lessons. Such challenges will be, certainly, implemented in the next reactor generation, following the technological evolution. The technical audits could be an important mechanism to identify the main procedures to be implemented in each operational NPP and its respective priority.

It is becoming clear that the continuity of the use of nuclear energy depends on the evolution of technologies and procedures improvement that should be incorporated to the emergency systems with appropriate redundancies, ensuring the adequate safety to the public, workers and environment. The efforts to provide such a condition must be of all stakeholders, including international organizations.

ACKNOWLEDGEMENTS

Zieli Dutra Thomé, Rogério dos Santos Gomes, Fernando Carvalho da Silva, Sergio de Oliveira Vellozo We thank to CNEN and FAPERJ as the financial sponsors and IME by the support facilities.

REFERENCES

1. Institute of Nuclear Power Operations (2011) Special Report on the Nuclear Accident at the Fukushima Daiichi Nuclear Power Station—INPO 011-005. INPO, Atlanta.

2. The National Diet of Japan (2012) The Official Report of the Fukushima Nuclear Accident Independent Investigation Commission. The National Diet of Japan, Tokyo.

3. Thomé, Z.D., Gomes, R.S., Silva, F.C. and Gomes, J.D.R.L. (2012) An Attempt to Confirm the Origin of ^{135}Xe Detected in the Fukushima Daiichi II Nuclear Power Plant. Nuclear Engineering and Design, 247, 123-127.

4. International Nuclear Safety Advisory Group (1992) The Chernobyl Accident: Updating of INSAG-1, INSAG 7. IAEA, Vienna.

5. International Nuclear Safety Advisory Group (1991) Safety Culture, INSAG 4. IAEA, Vienna.

6. International Atomic Energy Agency (2002) Self-Assessment of Safety Culture in Nuclear Installations—Highlights and Good Practices—TECDOC-1321. IAEA, Vienna.

7. International Atomic Energy Agency (2003) Flood Hazard for Nuclear Power Plants on Coastal and River Sites— Safety Standard Series NS-G-3.5. IAEA, Vienna.

8. International Atomic Energy Agency (2011) Meteorological and Hydrological Hazards in Site Evaluation for Nuclear Installations, Specific Safety Guide SSG-18. IAEA, Vienna.

9. International Atomic Energy Agency (2011) International Fact Finding Expert Mission of the Fukushima Dai-ichi NPP Accident Following the Great East Japan Earthquake and Tsunami, Mission Report. IAEA, Vienna.

10. Japan Society of Civil Engineers (2002) Tsunami Assessment Method for Nuclear Power Plants in Japan. JSCE, To- kyo.

11. Acton, J.M. and Hibbs, M. (2011) Why the Fukushima was Preventable. Carnegie Endowment for International Peace, Washington DC.

12. Suzuki, T. (2011) The Fukushima Nuclear Accident: Lessons Learned (So Far) and Possible Implications. Special Seminar on the Fukushima Dai-ichi Incident: Implications for the UK, France, Japan and the International Community, London, 6 July 2011.

13. Preckshot, G.G. (1994) Method for Performing Diversity and Defense-in-Depth Analyses of Reactor Protection Systems. U.S. Nuclear Regulatory Commission, Washington DC.

14. Hatamura, Y., Oike, K., et al. (2012) Investigation Committee on the Accident at Fukushima Nuclear Power Stations of Tokyo Electric Power Company.http://www.cas.go.jp/jp/seisaku/icanps/eng/final-report.html

15. Kaufmann, A., Grouchko, D. and Groun, R. (1977) Mathematical Models for the Study of the Reliability of Systems. Academic Press, New York.

16. Kelly, D.L. (1991) Overview of Containment Venting as an Accident Mitigation Strategy in US Light Water Reactors. Nuclear Engineering and Design, 131, 253-261.

17. Dallman, R.J. and Galyean, W.J. (1988) Containment Venting as an Accident Management Strategy for BWRs with Mark I Containments. Nuclear Engineering and Design, 121, 421-429.

18. United States Nuclear Regulatory (1989) Installation of a Hardened Wetwell Vent (Generic Letter 89-16). http://www.nrc.gov/reading-rm/doc-collections/gen-comm/gen-letters/1989/gl89016.html

19. Davies, L. (2011) Beyond Fukushima: Disasters, Nuclear Energy, and Energy Law. Brigham Young University Law Review, 2011, 1937-1989.

20. Greene, S.R. (1990) The Role of BWR Secondary Containments in Severe Accident Mitigation: Issues and Insights from Recent Analyses. Nuclear Engineering and Design, 120, 75-86.

21. Tokyo Electric Power Company (2012) Investigation of the Cause of Hydrogen Explosion at the Unit 4 Reactor Building.

22. Schlueter, R.O. and Schmitz, R.P. (1990) Filtered Vented Containments. Nuclear Engineering and Design, 120, 93- 103.

23. Organization for Economic Co-Operation and Development (1988) Filtered Containment Venting Systems. Note on the Outcome of the May 1988 Specialists Meeting on Filtered Containment Venting Systems, Paris, 17-18 May 1988.

24. Muhleim, M.D. and Wood, R.T. (2007) Design Strategies and Evaluation for Sharing Systems at Multi-Unit Plants, ORNL/LTR/INERI-BRAZIL/06-01. Oak Ridge National Laboratory, Oak Ridge.

25. US Nuclear Regulatory Commission (1975) Shared Emergency and Shutdown Electric System for Multi-Unit Nuclear Power Plants, Regulatory Guide 1.81. U.S. Nuclear Regulatory Commission, Washington DC.

26. US Nuclear Regulatory Commission (2003) Control Room Habitability at Light-Water Nuclear Power Reactors, Regulatory Guide 1.196. U.S. Nuclear Regulatory Commission, Washington DC.

27. Hirano, M., et al. (2012) Insights from Review and Analysis of the Fukushima Dai-Ichi Accident. Journal of Nuclear Science and Technology, 49, 1-17.

28. International Atomic Energy Agency (2009) Severe Accident Management Programmes for Nuclear Power Plants, IAEA Safety Guide NS-G-2.15. IAEA, Vienna.

29. Cook, D.H., Greene, S.R., Harrington, R.M., Hodge, S.A. and Yue, D.D. (1981) Station Blackout at Browns Ferry Unit One—Accident Sequence Analysis, NUREG/CR-2182. Oak Ridge National Laboratory, Oak Ridge.

30. International Atomic Energy Agency (2011) Mitigation of Hydrogen Hazards in Severe Accidents in Nuclear Power Plants, IAEA-TECDOC-1661. IAEA, Vienna.

31. Reineck, E., Tragsdorf, I.M. and Gierling, K. (2004) Studies on Innovative Hydrogen Recombiners as Safety Devices in the Containments of Light Water Reactors. Nuclear Engineering and Design, 230, 49-59.

32. Heck, R., Kelber, G., Schmidt, K. and Zimmer, H.J. (1995) Hydrogen Reduction Following Severe Accidents Using the Dual Recombiner-Igniter Concept. Nuclear Engineering and Design, 157, 311-319.

33. Bröckerhoff, P., von Lensa, W. and Reinecke, E. (2000) Innovative Devices for Hydrogen Removal. Nuclear Engineering and Design, 196, 307-314.

34. Electric Power Research Institute (2011) Silicon Carbide Provides Opportunity to Enhance Nuclear Fuel Safety.http://mydocs.epri.com/docs/CorporateDocuments/Newsletters/NUC/2011-09/09d.html

Chapter 10

A Longitudinal Study of Human Exposure to Potential Nuclear Power Plant Risk

Dean Kyne1, • Jason T. Harris2

1 Department of Sociology and Anthropology, The University of Texas Rio Grande Valley, Edinburg, TX 78539, USA

2 School of Health Sciences, Purdue University, West Lafayette, IN 47907, USA

ABSTRACT

This study constructs a potential risk index (PRI) for the 65 U.S.-based commercial nuclear power plant (NPP) sites in relation to their surrounding populations. Four risk levels are defined: low risk, moderate risk, high risk, and very high risk. Discrepancies that exist in the sociodemographic characteristics of the host communities' populations are examined as sorted by risk-level category. It is found that a greater percentage of minority groups are exposed to the highest levels of risk. In addition, percent "Hispanic" and percent "Other," a grouping that includes multiracial, mixed, interracial, as well as Hispanic and Latino groups (for example, Mexican, Puerto Rican, Cuban, or Spanish) are categories that show the greatest percent change in both the period 1990–2000 and 2000–2010.

KEYWORDS

Environmental justice, Nuclear power induced disaster, Nuclear power plant, Potential risk index

1. INTRODUCTION

The communities that host nuclear power plants (NPPs) inescapably face the various kinds of risks associated with them. In many instances, the NPP siting process was completed before concern for environmental justice became widespread, and the host communities did not have an opportunity to participate in the environmental decision-making process. Lack of opportunity to participate in decision making resulted in the NPP being built in areas in which the communities involuntarily shoulder the potential risks of negative environmental impacts that could stem from the plant at any time. As a result, communities are saddled with a number of environmental justice issues without having possible solutions.

This study provides answers to these research questions: what are the levels of potential risk associated with NPPs and what are the sociodemographic characteristics of specific populations living within a 50-mile radius of a NPP? Specifically, the authors investigate the levels of potential risk associated with NPPs and the sociodemographic characteristics of specific populations. To answer these questions, the authors first identify the potential generic risks associated with individual NPPs. The authors then present the sociodemographic characteristics of the populations living in NPP host communities within a 50-mile radius of a NPP, sorted into four levels of potential risk—low, moderate, high, and very high. After discussing trends that are evident in the demographics of these populations, the authors discuss the environmental justice issues these populations face and summarize their findings.

2. Potential Risks

The emergence of commercial nuclear power can be traced back to 8 December 1953, the day President Dwight Eisenhower gave his "Atoms for Peace" speech, in which he highlighted his plan to transform nuclear fission materials—then, as now, the subject of much public fear—into a resource with a peaceful end use (Eisenhower 1953). On 18 December 1957, Shippingport Atomic Power Station—the world's first commercial nuclear power plant, built on the Ohio River 25 miles northeast of Pittsburgh—began generating nuclear power (U.S. NRC 2011). As of 12 August 2012, there were 104 nuclear reactors located at 65 sites scattered across 31 states (U.S. NRC 2012a). The study included the 104 nuclear reactors in the data analysis. Four of these nuclear power plants were permanently shut down after 2012: they are Crystal River, Florida in February 2013; Kewaunee Carlton, Wisconsin in May 2013; San Onofre, California in June 2013; and Vermont Yankee, Vermont in December 2014 (U.S. NRC 2015).

In the 56 years since the reactors at Shippingport came online, NPPs have on occasion demonstrated a destructive potential that goes beyond what human capacity can manage. The first publically acknowledged NPP-related accident took place at Three Mile Island (TMI), 26 years after the Shippingport NPP came on line, when the plant encountered a major core-damage accident. Fortunately, the event came to an end on 28 March 1979 without requiring an evacuation. Residents living within a 5-mile radius of the plant, including preschool-aged children and pregnant women, were evacuated but only after 2 days had passed since the accident occurred (Three Mile Island Alert 2013).

The second demonstration of NPPs' powerful destructive potential occurred at Chernobyl NPP in Ukraine, seven years after the event at TMI. The Chernobyl plant experienced an accident that breached its reactor containment vessel and released radioactive effluents into the environment on 26 April 1986 (U.S. NRC 2013a). Areas within an 18-mile radius of the plant were abandoned and approximately 115,000 people were evacuated; an additional 220,000 people were evacuated in subsequent years. It was noted that the impacts were far felt beyond the 18-mile radius, reaching out to many areas of central and northern Europe (UNSCEAR2012).

Twenty-five years after the Chernobyl accident, the third and most recent NPP-related accident occurred when an underwater earthquake triggered a tsunami that hit the Japanese coastline, causing massive damage at the Fukushima Daiichi NPP (U.S. NRC 2013d). The Japanese government in its initial response evacuated approximately 160,000 people living within a 12- to 19-mile radius of the Fukushima plant (Morris-Suzuki et al. 2012), but the impacted areas have proven to be much wider than the evacuated areas.

These accidents at NPPs are normally preceded by an event or a series of events that trigger or contribute to the damage done to the reactor core. When brought under control, events with the potential to have escalated into major accidents—that is, those of the type capable of causing reactor-core damage—are described as near-miss events. The U.S. Nuclear Regulatory Commission (NRC) issues an annual report detailing all near-miss events and classifies them according to three categories defined by the type of team sent out to inspect the event, which is in turn determined by the level or degree of severity of the event. There are three such teams—namely, the augmented inspection team (AIT), the incident inspection team (IIT) and the special inspection team (SIT). The AIT evaluates events that pose a 10-fold increase in risk, whereas the SIT investigates events that pose a 1000-fold increase in risk level (Lochbaum 2011, 2012, 2013, 2014, 2015). In 2010, there were a total of 14 near-miss events (13 SIT and 1 AIT); in 2011, 15 (14 SIT and 1 AIT); in 2012, 14 (11 SIT and 3 AIT); in 2013, 10 (9 SIT and 1 AIT); and in 2014, 9 (9 SIT and 0 AIT). These figures hint at the regularity with which such plants pose a risk to their surrounding areas (Fig. 1). The total number of near-miss events coupled with a decline in total number of reactors showed a decline trend in the past two years.

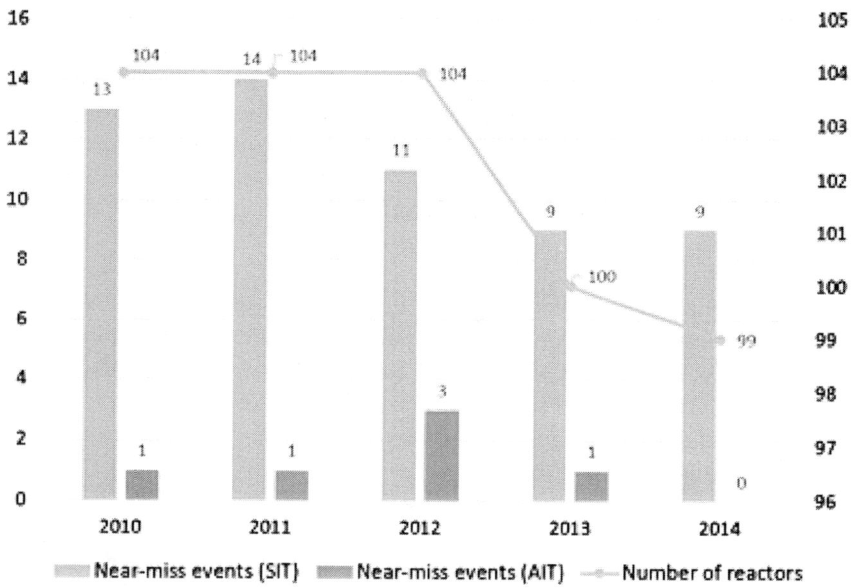

Figure 1. Near-miss events taking place at U.S. commercial nuclear power reactors. *Source* Lochbaum (2011, 2012,2013, 2014, 2015)

Another risk associated with NPPs is that of contamination of the sort caused by the unmonitored and unplanned release of liquids. Commercial NPPs release radioactive materials into the environment, in either or both liquid or gaseous form, on a routine basis. There have been incidents of unplanned and unmonitored leaks of liquids that occurred at the Braidwood, Indian Point, Byron, and Dresden NPPs. The Liquid Radioactive Release Lessons Learned Task Force (LLTF) has been tasked with identifying the causes of such leaks (Richards et al. 2011). Their findings included the following: (1) the construction of plant components did not meet existing safety standards; (2) the components that caused or contributed to said leaks were not required to be monitored, nor were they subject to routine maintenance activities according to NRC regulations; (3) some components associated with subterranean leaks or with spent-fuel pools were not readily accessible or physically visible; (4) liquid leaks can enter undetected into groundwater; (5) the contamination of groundwater may go undetected because it is not required to be monitored according to existing NRC regulations; and (6) the contaminated groundwater could flow offsite undetected. The communities that host NPPs therefore, are at perpetual risk due to the presence of their local NPP.

A different potential risk associated with NPPs is their spent fuel—the highly radioactive used fuel rods that are removed from a nuclear reactor. Nuclear fuel rods are made of fissionable materials and retain their inherent destructive potential long after their initial use. Approximately 74 % of all spent fuel is stored at the reactor site, most often in spent-fuel pools, while the rest are stored in dry-storage casks distributed across 33 states (U.S. GAO 2012). The U.S. Government Accountability Office (GAO) projects an annual 2000-metric-ton increase in the amount of spent fuel stored on site, with the current 70,000-metric-ton spent-fuel stockpile growing to 140,000 metric tons within a decade. New storage facilities are expected to be ready to begin accepting spent fuel in 15–40 years. The potential risks associated with such facilities include: (1) the release of radiation, which could have severe negative impacts on human health; and (2) a self-sustaining fire if and when the water is drained and the fuel rods are exposed to the air. Above all, the threats that are inherent to the 70,000-metric-ton spent-fuel stockpile cannot and must not be underestimated.

This spent fuel is also always vulnerable to terrorist attack (Holt and Andrews 2007). The terrorist attack on the World Trade Center in New York and the Pentagon building in Washington, DC on 11 September 2011—the so-called 9/11 attack—serves as reminder of the dire consequences and very real threat of a potential attack on a NPP. Afterwards, evidence came to light that revealed Al Qaeda had considered targeting a NPP in their initial plan of attack (Holt and Andrews 2007).

According to the National Academy of Sciences (U.S. NAS 2006), a successful attack on a spent-fuel storage facility would be difficult but is possible. In the event of such an attack, the spent fuel can become the source of a self-sustaining zirconium cladding fire, and will in such an event release a massive amount of radioactive materials. Communities that host NPPs therefore live with the unpredictable risk of exposure to a massive release of radiation.

The intent is that these materials will not permanently reside in the spent-fuel pools or stay safely in the dry-storage casks housed at the reactor sites. When they are transported, they pose risk not only to the host communities but also to the communities at large along the transit route. The purpose of transporting spent fuels is to relocate them to storage space shared by multiple reactor sites operated by the same owner (U.S. NRC 2013b). According to Garrick (2003), between 1964 and 1997, a total of 3025 shipments moved 829 metric tons of heavy metal (MTHM) by road and an additional 1445 MTHM by rail. The transportation of such materials is closely supervised by the NRC and the U.S. Department of Transportation (U.S. NRC 2013b). Although regulated and supervised, the transportation—whether by road or by rail—has not been accident free. Between 1971 and 1995, there were four accidents on the highways and four on the rail lines; one highway accident resulted in the death of a driver and detectable emissions of radiation (Garrick 2003; U.S. NRC 2013e). All of the approximately 70,000 metric tons of spent fuel stored at reactor sites are waiting for shipment to high-level repositories, when such facilities are open and available to accept nuclear waste materials (U.S. NRC 2013b). With the Obama administration's 2009 decision to withdraw funding for the Yucca Mountain High-Level Nuclear Waste project, there is at present no permanent storage facility under development (Wald 2009).

Another risk that NPPs pose is inherent to their design and serves to highlight why it is critical that the design of a reactor that contains fissionable fuel rods during its normal operation be centered on safety. The Mark 1 containment vessel, first produced by GE in the 1960s and still in use at Fukushima NPP at the time of the accident there, contributed to the catastrophic events of 2011 (Zeller 2011). As early as 1975, it was known that there had not been sufficient testing of the containment design and that any flaws that might persist could compromise the safety of the plant and its surroundings. But the warning did not lead to a halt in the operating of the design because it had by that time been widely accepted throughout the nuclear power industry. Harold Denton, a retired NRC official, pointed out that the probability of a Mark 1-type reactor bursting if the fuel rods were to enter into meltdown was approximately 90 % (Denton 1987). In the United States, there remain 23 Mark 1 reactors that are exactly the same reactor design as those four reactors damaged at the Fukushima NPP site still in operation, located at 16 NPPs, including: Oyster Creek, New Jersey; Dresden, Illinois; and Monticello, Minnesota. The host communities of NPPs that still operate Mark 1 reactors are at greater risk of exposure to a Fukushima-like disaster than sites with safer reactor designs. Nuclear power plants are also vulnerable to seismic events of the type that played a role in triggering the catastrophe at Fukushima. The NPPs sited along the U.S. eastern seaboard in particular do not include in their designs any kind of measures meant to shield against the aftereffects of a strong earthquake (Koch 2011).

At the core of many of these concerns regarding NPPs is the fact that radiation poses a risk to public health. Radiation is a public health concern for two main reasons. First, humans' ability to structure their environment, as individuals, is rapidly decreasing because of population implosion. This results in the inability to completely avoid exposure from human-made sources of radiation. Second, radiation health effects are not unique and are detected only statistically. Standards and regulations for protection must be developed using the information provided by the scientific and medical communities with input from the public and government. Additionally, there are two other aspects of radiation exposure that need to be understood to properly evaluate the risk to public health. First, not all people exposed to radiation are affected equally. For example, children are much more vulnerable than adults to the effects of radiation. Second, radiation from internal emitters is in a sense different from external beam radiation. Irradiation by external beam radiation exists only as long as the source is present. Irradiation stops if the source is removed from the vicinity of the individual or the individual is removed from the vicinity of the source. However, radioactive materials taken into the body (that is via inhalation or ingestion) will continue to irradiate the individual for as long as it is in the body and the material remains radioactive.

Both the International Commission on Radiological Protection (ICRP 2007) and the U.S. National Academy of Sciences (U.S. NAS/NRC 2006) concluded that any exposure, including exposure to naturally occurring background radiation, creates an increased risk of cancer. This conclusion follows the linear non-threshold (LNT) model. Specifically, the LNT model assumes that, in the low dose range, radiation doses greater than zero will increase the risk of excess cancer and/or heritable disease in a simple proportionate manner. While further research is needed to undoubtedly prove this is true, current findings support this model and it has been adopted in standards and regulations worldwide. Regardless of whether or not the source of radiation is human-made, as in the case of NPPs, or natural, as exposure increases so does risk.

The communities that host such facilities shoulder the potential risks associated with nuclear power from the time the plant begins operation until the time it is decommissioned—for a typical reactor design, the use life is between 40 and 60 years (U.S. NRC 2014). After a plant is shutdown, it takes under normal conditions 50 years for the reactor to cool; this is followed by a period of decommissioning that normally takes another 60 years. For example, Crystal River Unit 3 was permanently shut down on 2 February 2013 and NRC scheduled to complete decommission in 2073 (U.S. NRC 2014). The NRC has advanced three strategies for dealing with end of use at a NPP: (1) DECON, or decommissioning strategies; (2) SAFSTOR, or deferred dismantling; and (3) ENTOMB, or the permanent encasing on site of radioactive contaminants (U.S. NRC 2013c). For example, Hallam Nuclear Power Plant, Nebraska, a NPP with a short use life, was decommissioned in 1971. All potential containments were at that time buried underground. Since then, the Department of Energy, working in conjunction with the Nebraska Department of Health, has monitored the entombment site for possible groundwater contamination and radiation, and will continue to do so for a period of 119 years, until 2090 (Nebraska Energy Quarterly 1997). Figure 2 shows the location of the permanently decommissioned U.S.-based commercial nuclear power reactors as of August 2012.

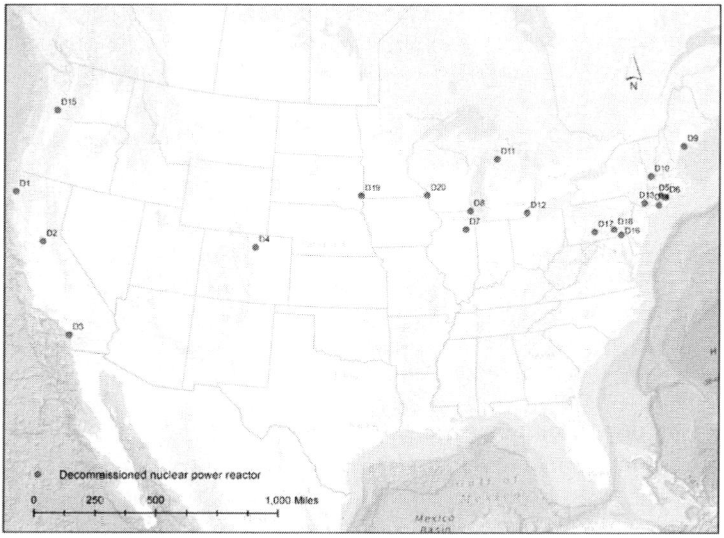

Index	Plant Name, Reactor	
D1	Humboldt Bay Nuclear Power Plant, Humboldt Bay 3	
D2	Rancho Seco Nuclear Generating Station, Rancho Seco	
D3	San Onofre Nuclear Generating Station, Unit 1	*
D4	Fort Saint Vrain Generating Station, Fort St. Vrain	
D5	Connecticut Yankee, Haddam Neck	*
D6	Millstone Power Station, Unit 1	*
D7	Dresden Nuclear Power Station, Unit 1	
D8	Zion Nuclear Power Station, Zion 1 & 2	#
D9	Maine Yankee Nuclear Power Plant, Maine Yankee	
D10	Yankee-Rowe Nuclear Power Station, Yankee-Rowe	
D11	Big Rock Point Nuclear Plant (Monticello)	
D12	Fermi, Unit 1	
D13	Indian Point Nuclear Generating Station, Unit 1	*
D14	Shoreham Nuclear Power Station, Shoreham	*
D15	Trojan Nuclear Power Plant, Trojan	
D16	Peach Bottom Atomic Power Station, Unit 1	
D17	Saxton	*
D18	Three Mile Island Nuclear Station, Unit 2	
D19	Pathfinder Atomic Plant, Pathfinder	*
D20	La Crosse Nuclear Generating Station, La Crosse	

Figure 2. Permanently decommissioned U.S.-based commercial nuclear power reactors as of August 2012. *Data source* U.S. NRC (2012b). *Note* *Located in currently operating nuclear power plant site. #Entire plant site with the two reactors was permanently shut down

Above all, the potential risks associated with NPPs can be explained by Perrow's normal accidents theory. As noted previously, NPPs are complex and tightly coupled systems and complexity produces unknown risks. According to Perrow's theory:

Nothing is perfect, neither designs, equipment, procedures, operators, supplies, or the environment. Because we know this, we load our complex systems with safety devices in the form of buffers, redundancies, circuit breakers, alarms, bells, and whistles. Small failures go on continuously in the system since nothing is perfect, but the safety devices and the cunning of designers, and the wit and experience of the operating personnel, cope with them. Occasionally, however, two or more failures, none of them devastating in themselves in isolation, come together in unexpected ways and defeat the safety devices—the definition of a "normal accident" or system accident. If the system is also tightly coupled, these failures can cascade faster than any safety device or operator can cope with them... if the accident brings down a significant part of the system, and the system has catastrophic potential, we will have a catastrophe. (Perrow 1999, pp. 356–357)

According to Perrow's analysis, the risks associated with NPPs therefore should be seen as intrinsic to the system, normal, and as something that cannot be avoided.

3. METHOD

This section discusses the study area, variables, data, method, and construction of the NPP PRI index.

3.1 Study Area

This study focuses on two distance areas surrounding the 65 NPPs. The two distance areas are similar to those identified in the recent study of Kyne (2015). One is those areas within a 50-mile radius of any one of the 65 U.S.-based NPPs; the other is those areas outside of a 50-mile radius and yet that are located in the state(s) that fall within said 50-mile radius. There are 20 permanently deactivated reactors located at NPPs that otherwise remain active. Kyne (2015) provided a notion that these 20 reactors do not pose the same level of risk as do reactors currently in operation, which pose increased risks as a result of their day-to-day operation, the possibility of core meltdown-type accident, and from their release of gaseous and liquid radioactive effluents. However, they could be regarded as a potential source of radiation risk from any spent fuel stored on site. The reason for exclusion was to ensure apples to apples comparison.

3.2 Study Variables

The study includes variables that are similar to those used in the study of Kyne (2015). The racial/ethnic variables included the categories (1) White—Percent White as percent of all non-Hispanic Whites; (2) Black—Percent Black as percent of non-Hispanic Blacks or African Americans; (3) Asian—Percent Asian as percent of Asian, and Native Hawaiian or Other Pacific Islander; (4) Hispanic—Percent Hispanic as percent of people who are Hispanic origin; (5) Native American—Percent Native American as percent of American Indian or Alaska Native; (6) Other—Percent Other as percent of Some Other Race who are not included in the White, Black, or African American, American Indian or Alaska Native, Asian, and Native Hawaiian or Other Pacific Islander race categories, or whose identify themselves as multiracial, mixed, interracial, or a Hispanic or Latino group (for example, Mexican, Puerto Rican, Cuban, or Spanish); and (7) Color—Percent Color as percent of all other races except non-Hispanic Whites. Similarly, the study included the socioeconomic variables that were also used in the study of Kyne (2015), which are renter-occupied housing, unemployment rate, percent living in poverty, mean household income, and percent composition at the census-tract level.

Another study variable was distance, which was measured as Euclidean distance in miles from the census-tract center point to the NPP using the ENVI's ArcMap program (Kyne 2015). The distance was used to investigate whether and how disparities in certain demographic characteristics were associated with the distance from a NPP.

Other study variables relate to radioactive materials and radiation dosage levels. These include: radioactive effluents released from a reactor at a given NPP; the radiation dose able to be delivered by those effluents; and the amount of radiation absorbed by the human body upon exposure.

3.3 Study Data

This study included multiple datasets. To investigate demography-based disparities in the study areas, this study required three discrete datasets:

1 First, for demographic characteristics, data were obtained from the United States Census Bureau and GeoLytics (GeoLytics 2012a, 2012b, 2012c), which were also used in the previous study (Kyne 2015). The census tract level data allowed for comparison in the study years because the census tract boundaries are normalized to those of the 2010 data.

2 Similar to the datasets used in the study of Kyne (2015), this study includes the 2010 Census TIGER/Line Shapefiles and the 2010 U.S. Census Summary File 1 Demographic Profile (DP1) for the United States and Puerto Rico were used in order to join the normalized data containing study variables (United States Census Bureau 2010). The 2010 Census TIGER/Line Shapefiles were purposefully used because the most recent decadal dataset used in this study was 2010. In addition, the data in 1990 and 2000 were normalized to those of the 2010 boundaries.

3 The location of each of the 104 nuclear reactors located at 65 sites that are currently in operation in the United States was obtained from the NRC's website (U.S. NRC 2012b). As noted earlier, there are four power plants that were permanently shut down before the time this manuscript is published. It results in a fleet of 99 reactors located in 61 sites across 30 states as of August 2015 (U.S. NRC 2015).

To study the extent and severity of exposure to radiation and radioactive effluents, our research required access to two datasets:

1 Radioactive effluents data from 2001 to 2008 were acquired via the Center for Advanced Energy Studies (Harris2013). The effluents data include both liquid and gaseous data for both pressurized water reactors (PWRs) and boiling water reactors (BWR).

2 The reactor oversight dataset includes seven variables that detail three strategic areas—reactor safety, radiation safety, and safeguards. The seven variables are: initiating events, mitigating systems, barrier integrity, emergency preparedness, public radiation safety, occupational radiation safety, and physical protection. These have been published in *1Q/2012 ROP Performance Indicators Summary* (U.S. NRC 2012a).

To examine the risks of natural disaster associated with individual NPPs, data regarding earthquakes (FEMA 2013; USGS 2013), hurricanes, tornadoes, and volcanoes (The Daily Beast 2011) were also obtained. These data contain earthquake information measured on a seven-point scale based on the United States Geological Survey (USGS) map; hurricane predictions based on historical data as compiled by the USGS; the number of predicted tornado days per century based on a dataset spanning 1921–1995 (as complete and predictive a measure of tornadoes as is currently available); and NPPs sorted by proximity to an active volcano.

3.4 Study Method

This study mainly utilized two research methods in searching answers to the research question. First, the study used center point-distance suggested in the study of Kyne (2015) to measure distance from the NPPs to the hosting communities that were located within a 50-mile radius from the nearest NPP. Measuring the distance included two steps. Step one included determining the geographic center point of each included census tract in ArcMap 10.1 and next selecting those geographic center points that fell within a 50-mile radius of the nearest NPP, identified using the “near” tool in ArcMap 10.1. This method provided the exact distance between the census tracts’ center points and the relevant NPPs. Step two included collapsing the demographic characteristics of population living in the census tracts within a 50-mile radius. Second, the study constructed a potential risk index (PRI), which is a composite index that can be applied to all U.S.-based NPPs. The composite index measures the potential risks associated with NPPs. According to the OECD (2008), “a composite indicator is formed when individual indicators are compiled into a single index on the basis of an underlying model of the multi-dimensional concept that is being measured” (OECD 2008, p. 13). To construct the index, a five-step process was followed: (1) construct a theoretical framework; (2) select data; (3) normalize the data; (4) determine the weights to be assigned to the different categories and types of data; and (5) aggregate the data.

3.5 NPP Potential Risk Index (NPP PRI)

A conceptual framework was initially constructed that includes the three cornerstone potential risks that confront any NPP—plant operational risks, natural disaster risks, and plant locational risk (Fig. 3). The potential risks at any NPP can stem from the plant itself—from its day-to-day operations, the normal aging of the plant, and/or the continued, ongoing exposure of the plant, its equipment and its surroundings to radioactive effluents and their byproducts. Next, any NPP could experience an accident, including a core-damage event, which could be triggered by a natural disaster such as an earthquake, hurricane, or tornado. The plant’s physical location, especially when near to densely populated areas, influences the level of the potential risk it poses. If a given plant is sited in proximity to a densely populated area, then the plant would be potentially exposed to human-caused hazards such as an airplane crash or a terrorist attack. The potential risk level of each NPP can be measured across these three dimensions.

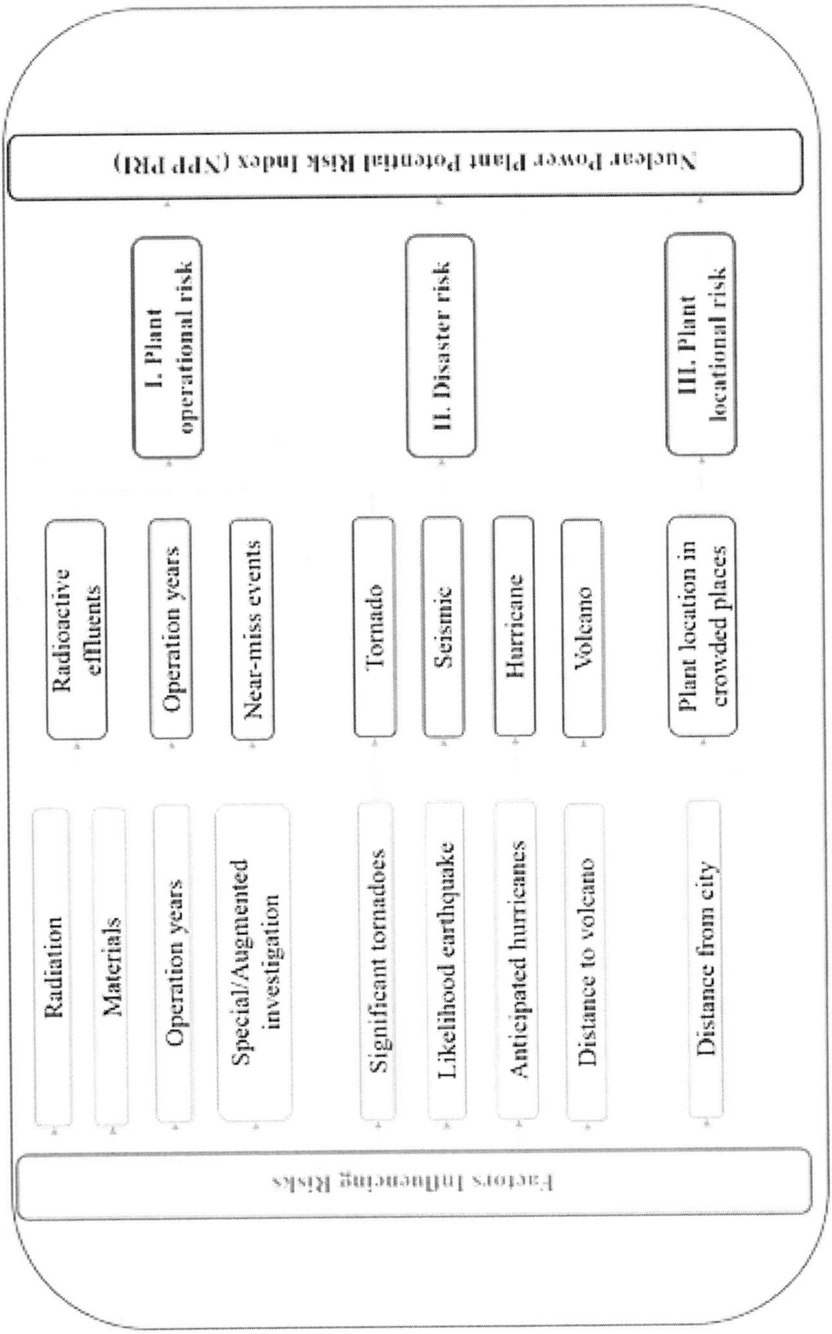

Figure 3. A conceptual framework used in constructing a PRI for U.S.-based NPPs

Radioactive effluents are measured by the radioactive materials that comprise the waste stream, and the associated radiation doses those materials deliver or carry with them. Data for two types of radioactive effluents—gaseous and liquid effluents—were summed to arrive at a total volume of radioactive effluents released by each NPP. The amount released was normalized according to the amount of electricity generated at each site. The effective dose data, which were computed using the normalized total radioactive effluents, were then estimated using a dose estimate model established by the United Nations Scientific Committee on the Effects of Atomic Radiation (UNSCEAR) and employed by Harris and Miller (2008). The dose estimate model made an assumption that a model NPP site has a population density of 1036/square mile and 52/square mile for within 31 miles of the site and within 772 miles of the site, respectively (Kyne 2015). This model site provides an estimate of the collective effective dose (CED) per unit release for different release categories. The CED per unit of electrical energy (man Sv (GW a) – ^1) then can be obtained by multiplying the normalized releases by the UNSCEAR-calculated values for CED per unit release. Both the radioactive effluents released and their effective dose data for the period 2001–2008 were included in these calculations, and were given equal weight in constructing the NPP PRI presented here.

Another type of data is near-miss events. The NRC typically reports on near-miss events at NPPs. These events are termed "near miss" because they raise the risk of damage to the reactor core and endanger the safety of workers and the public, but do not result in core meltdown or radioactive effluent release (Lochbaum 2011,2012, 2013, 2014, 2015). Analysis of a near-miss event will, depending on the potential risk level of the event, involve one of the NRC's three investigative teams: the augmented inspection team; the incident inspection team; or the special inspection team. In the past 3 years, there have been reports of near-miss events issued by each of the first two teams. Known near-miss events that occurred in 2010, 2011, and 2012 were accounted for in constructing the NPP PRI when the data analyses were conducted in August 2013. According to the World Nuclear Association (2013), most NPPs are designed for a use life of 30–40 years. As this window is extended out beyond planned operational life, NPPs will face three problems. First, there are problems at aging plants with components and parts that need to be replaced, some of which are difficult to access. Second, there are issues with obsolescence, as many NPPs currently in operation still use analog instrument and control systems. Third, due to heat and neutron irradiation, a plant's materials degrade with age. In the United States, World Nuclear Association (2013) anticipated that more than 100 reactors will be granted license extensions from 40 to 60 years. These extensions pose a safety risk and raise questions about the reliability at older NPPs. The age of the plant, or its duration of operation in years, can be computed by subtracting the year of issue of its commercial operating license from the current year. The longer a NPP has been in operation, the more likely it is that it may pose an increasing risk.

A natural disaster such as an earthquake, hurricane, or tornado may trigger a nuclear accident that can result in damage being done to the reactor core. The likelihood of an earthquake striking a NPP was estimated with data taken from the United States Geologic Survey's (USGS) National Seismic Hazard Map. This map expresses the severity of horizontal shaking as a percentage of g—the acceleration of a falling object due to gravity—and displays shaking intensity with seven color-coded categories that denote "seismic design categories" (USGS 2014). The categories correspond to the likelihood of an earthquake occurring that would lead to shaking and displacement of various intensities (FEMA 2013). The USGS map was overlaid on a map showing the locations of all 65 NPPs in the United States. The output display was then used to assign each NPP a value for the likelihood that an earthquake will occur, on a seven-point scale ranging from 0 (least likely) to 6 (most likely). Other natural hazards were considered less likely to generate high risk levels, but were included in the index analysis for completeness and because of their importance for particular NPP sites. Hurricane risk was measured by taking into account the number of hurricanes predicted to make landfall near the site of a NPP in the century to come (The Daily Beast 2011). Tornado risk was measured by considering the average number of significant tornadoes documented between 1921 and 1995 (The Daily Beast 2011). Volcano risk was measured by distance miles from the nearest volcano (The Daily Beast 2011). Finally, locational risk was measured by distance in miles from the nearest city.

Selected study indicators were normalized in order to render them comparable to one another. The min–max method was used to create an identical range by transforming indicators to a scale that would allow for direct comparisons. The method called for subtracting the minimum value from the original value and then dividing by the range of the indicator values. In other words, each variable was rescaled to a range from 0 to 100, using the following formula: $c_i' = [(c_i - c_{min})/(c_{max} - c_{min})] \times 100$, where c_i' = normalized or rescaled value; c_i = original value; c_{min} = minimum value in the variable; and c_{max} = maximum value in the variable (OECD 2008; Tate et al. 2010).

Construction of a composite index requires the application of appropriate weighting methods. Principal component analysis (PCA) and factor analysis (FA) methods were applied to aid in selecting appropriate weights, which were then used to compile the selected study indicators into a single composite index. Factor analysis with varimax rotation was then performed in order to minimize subindicators that had a loading on the same factor, which allowed us to study that subset of the principal components that accounted for the largest amount of variance. Four factors that were associated with eigenvalues larger than one were selected; these factors were responsible for a cumulative contribution to the overall variance of 63.19 % (Table 1). Weights from the factor loading results then were computed. To do so, the weights first were normalized by the squared factor loading (OECD 2008; Tate et al. 2010). For example, radiation dose in factor 4, of 0.863 loading, was normalized to 0.608, according to the formula 0.608 = (0.863^2)/1.2238. After normalizing the factor loadings, the composite weights were obtained. Following the method Nicoletti et al. (2000) prescribe, four subindicators were grouped with the highest factor loadings into four composite indicators: (1) subindicator 1 includes significant tornados (with a weight of 0.197), likelihood of earthquake (weight = 0.348), and distance from volcano (weight = 0.383); (2) subindicator 2 contains radioactive effluents (weight = 0.398), years in operation (weight = 0.359), and near-miss events (weight = 0.040); (3) subindicator 3 comprises anticipated hurricanes (weight = 0.508) and plant location distance from city (weight = 0.267); and (4) subindicator 4 is constituted by radiation dose (weight = 0.863). Next each of the four immediate subindicators was aggregated by applying the weight of the proportion of the explained variance in the dataset: 0.2853 for subindicator 1 (0.2853 = 1.6222/(1.62 + 1.47 + 1.37 + 1.22)); 0.2579 for subindicator 2; 0.2416 for subindicator 3; and 0.2152 for subindicator 4. The NPP PRI was obtained via the following equation: NPP PRI = (0.2853 × subindicator 1 + 0.2579 × subindicator 2 + 0.2416 × subindicator 3 + 0.2152 × subindicator 4).

To understand the relative risk levels that exist among the 65 NPPs operating in the United States, the composite index scores were ranked from highest to lowest. These were grouped into four categories based on their percentile rank: (1) low risk, between 0 and 25th percentile; (2) moderate risk, between >25th and 50th percentile; (3) high risk, between >50th and 75th percentile; and (4) very high risk, between >75th and 100th percentile. There are 16 plants in each of the categories low risk, moderate risk, and high risk, and 17 in the category very high risk (Table 2; Fig. 4).

4. FINDINGS AND DISCUSSION

This section discusses sociodemographic compositions by NPP PRI level and changes in demographic compositions during the study years. First, the study provides a discussion on findings of demographic characteristics of the population living within a 50-mile radius and their respective exposure to different levels of NPP PRI in 2010. Second, the study traces changes of demographic characteristics between the two study periods, 1990–2000 and 2000–2010, and looks into changes in demographic characteristics associated with the four levels of NPP PRI in the two study periods.

Table 1. Factor loadings of the NPP PRI using the varimax method

Variable	Factor loadings				Normalized factor loadings			
	Factor 1	Factor 2	Factor 3	Factor 4	Factor 1	Factor 2	Factor 3	Factor 4
Radiation dose	−0.010	0.096	−0.068	*0.863*	0.000	0.006	0.003	*0.608*
Radioactive effluents	−0.121	*0.764*	0.087	0.306	0.009	*0.398*	0.006	0.077
Years in operations	−0.282	*−0.725*	0.112	0.186	0.049	*0.359*	0.009	0.028
Near-miss events	0.068	*0.242*	0.189	−0.116	0.003	*0.040*	0.026	0.011
Significant tornados	*−0.565*	0.348	0.477	−0.330	*0.197*	0.083	0.166	0.089
Likelihood of earthquake	*0.751*	0.186	0.097	0.020	*0.348*	0.024	0.007	0.000
Anticipated hurricanes	−0.131	−0.055	*−0.835*	0.117	0.011	0.002	*0.508*	0.011
Plant location distance from city	−0.026	−0.360	*0.605*	0.448	0.000	0.088	*0.267*	0.164
Distance from volcano	*0.789*	0.015	0.112	−0.121	*0.383*	0.000	0.009	0.012
Eigenvalues	1.6222	1.4668	1.3742	1.2238				
Explained variance/total	0.2853	0.2579	0.2416	0.2152				

Italicized values indicate strong association between the variables and the latent factors. Factor 1, 2, 3, and 4

Table 2. PRI and PRI categories for the U.S.-based NPPs

No.	Nuclear power plant potential risk index (NPP PRI)	Risk category	Plant name	State	Index number in map (Fig. 4)	Date NPP permanently shut down	Number of mark I type reactor unit
1	42.49	Very high risk	Millstone Power Station	Connecticut	7		
2	32.17	Very high risk	Brunswick Steam Electric Plant	North Carolina	42		2
3	28.20	Very high risk	Joseph M. Farley Nuclear Plant	Alabama	2		
4	26.87	Very high risk	Surry Nuclear Power Station	Virginia	62		
5	26.63	Very high risk	Diablo Canyon Nuclear Power Plant	California	5		
6	26.14	Very high risk	Edwin I. Hatch Nuclear Plant	Georgia	11		2
7	25.95	Very high risk	H. B. Robinson Steam Electric Plant	South Carolina	54		
8	25.84	Very high risk	Turkey Point Nuclear Generating Station	Florida	10		
9	25.76	Very high risk	Waterford Steam Electric Station	Louisiana	22		
10	25.67	Very high risk	Oconee Nuclear Station	South Carolina	53		
11	25.54	Very high risk	Vogtle Electric Generating Plant	Georgia	12		
12	24.66	Very high risk	St. Lucie Plant	Florida	9		
13	24.56	Very high risk	Vermont Yankee Nuclear Power Plant	Vermont	60	12/29/2014	
14	24.47	Very high risk	Virgil C. Summer Nuclear Station	South Carolina	55		
15	24.45	Very high risk	LaSalle County Station	Illinois	17		
16	24.42	Very high risk	Peach Bottom Atomic Power Station	Pennsylvania	49		2
17	23.76	Very high risk	Grand Gulf Nuclear Station	Mississippi	30		
18	23.71	High risk	McGuire Nuclear Station	North Carolina	43		
19	23.67	High risk	Arkansas Nuclear One	Arkansas	4		
20	23.62	High risk	Columbia Generating Station	Washington	63		
21	23.24	High risk	Salem Nuclear Generating Station	New Jersey	37		
22	23.10	High risk	Pilgrim Nuclear Power Station	Massachusetts	24		1
23	22.90	High risk	Calvert Cliffs Nuclear Power Plant	Maryland	23		
24	22.76	High risk	North Anna Power Station	Virginia	61		
25	22.68	High risk	Three Mile Island Nuclear Station	Pennsylvania	51		
26	22.18	High risk	Indian Point Nuclear Generating Station	New York	40		
27	21.93	High risk	Sequoyah Nuclear Plant	Tennessee	56		
28	21.87	High risk	San Onofre Nuclear Generating Station	California	6	6/12/2013	
29	21.70	High risk	Fort Calhoun Station	Nebraska	33		
30	21.69	High risk	River Bend Station	Louisiana	21		
31	21.49	High risk	Catawba Nuclear Station	South Carolina	52		
32	21.33	High risk	Callaway Plant	Missouri	31		
33	20.94	High risk	Shearon Harris Nuclear Power Plant	North Carolina	44		

Table 2 continued

No.	Nuclear power plant potential risk index (NPP PRI)	Risk category	Plant name	State	Index number in map(Fig. 4)	Date NPP permanently shut down	Number of mark I type reactor unit
34	20.51	Moderate risk	Crystal River Nuclear Generating Plant	Florida	8	2/20/2013	
35	20.29	Moderate risk	Comanche Peak Steam Electric Station	Texas	58		
36	20.23	Moderate risk	Donald C. Cook Nuclear Power Plant	Michigan	25		
37	20.10	Moderate risk	Wolf Creek Generating Station	Kansas	20		
38	19.73	Moderate risk	Dresden Nuclear Power Station	Illinois	16		2
39	19.66	Moderate risk	Browns Ferry Nuclear Plant	Alabama	1		3
40	19.45	Moderate risk	Palisades Nuclear Plant	Michigan	27		
41	18.98	Moderate risk	Seabrook Station	New Hampshire	34		
42	18.60	Moderate risk	Byron Station	Illinois	14		
43	18.52	Moderate risk	Limerick Generating Station	Pennsylvania	48		
44	18.47	Moderate risk	R.E. Ginna Nuclear Power Plant	New York	39		
45	18.40	Moderate risk	South Texas Project	Texas	59		
46	18.36	Moderate risk	Oyster Creek Nuclear Generating Station	New Jersey	36		1
47	18.27	Moderate risk	Quad Cities Nuclear Power Station	Illinois	18		2
48	18.20	Moderate risk	Monticello Nuclear Generating Plant	Minnesota	28		1
49	17.91	Moderate risk	Davis-Besse Nuclear Power Station	Ohio	45		
50	17.69	Low risk	James A. FitzPatrick Nuclear Power Plant	New York	38		1
51	17.52	Low risk	Beaver Valley Power Station	Pennsylvania	47		
52	17.38	Low risk	Palo Verde Nuclear Generating Station	Arizona	3		
53	17.34	Low risk	Point Beach Nuclear Plant	Wisconsin	65		
54	17.26	Low risk	Hope Creek Generating Station	New Jersey	35		1
55	17.21	Low risk	Nine Mile Point Nuclear Station	New York	41		1
56	16.99	Low risk	Watts Bar Nuclear Plant	Tennessee	57		
57	16.84	Low risk	Cooper Nuclear Station	Nebraska	32		1
58	16.82	Low risk	Duane Arnold Energy Center	Iowa	19		1
59	16.07	Low risk	Braidwood Station	Illinois	13		
60	15.70	Low risk	Prairie Island Nuclear Generating Plant	Minnesota	29		
61	14.95	Low risk	Kewaunee Power Station	Wisconsin	64	5/7/2013	
62	14.22	Low risk	Susquehanna Steam Electric Station	Pennsylvania	50		
63	13.67	Low risk	Clinton Power Station	Illinois	15		
64	11.64	Low risk	Perry Nuclear Power Plant	Ohio	46		
65	11.50	Low risk	Fermi	Michigan	26		1

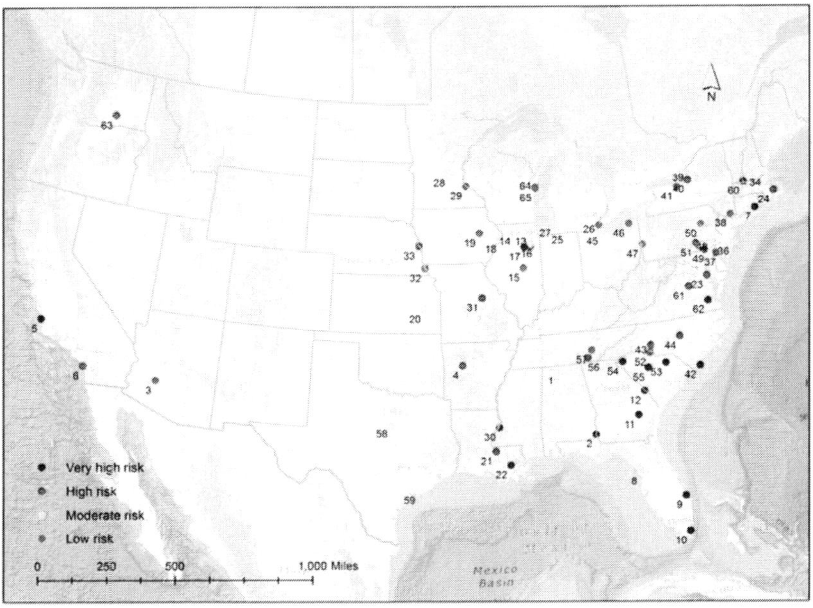

Figurre 4. Distribution of NPP-related potential risk in the United States

4.1 The Sociodemographic Composition of Populations According to NPP PRI Level

The 17 NPPs classified as very high risk accounted for the largest geographic area, some 83,640 square miles, or 28 % of the total land area occupied by NPPs in the United States. The highest percent Black (22 %) among potentially at-risk populations was found in the host communities of NPPs sorted into this subgroup. The second largest area, 74,961 square miles or 25 % of the total area occupied by NPPs in the United States, was associated with the 16 NPPs assigned to the high-risk category of the PRI as of 2010 (Table 3; Fig. 5). Because the census tract boundaries for prior-year data were normalized to the 2010 census-tract boundaries, these areas do not change when analyzing U.S. Census data for the years 1990 and 2000.

The at-risk populations included a greater percent Asian, percent Hispanic, percent Native American, percent Other, and percent Color than compared to their counterpart populations living in communities outside of the areas prone to nuclear power-related risks for each of the four PRI categories (Table 3). Populations living in close proximity to a NPP classified as low risk registered the highest percent unemployed and the highest percent living in renter-occupied housing units. Similar patterns were identified in the 1990 and 2000 U.S. Census data (Fig. 4).

Table 3. Demographic characteristics of populations living within a 50-mile radius of a NPP, classified according to four PRI categories using the 2010 U.S. Census data

2010	Low risk	Moderate risk	High risk	Very high risk	Total	Outside
Tracts	8445	5659	4966	4325	23,395	49,662
Tract area (sq. mile)	71,490	70,709	74,961	83,640	300,801	3,495,942
Total population	31,989,280	24,130,408	22,277,072	17,674,888	96,071,648	207,893,616
White	21,626,292	18,518,612	15,479,203	12,423,962	68,048,072	156,847,632
Black	5,402,879	2,861,589	3,268,325	3,949,417	15,482,210	22,496,542
Asian	2,036,120	1,041,474	1,398,483	399,936	4,876,013	9,801,153
Native American	109,554	67,366	95,330	60,454	332,704	2,147,761
Other	2,814,435	1,641,369	2,035,730	841,119	7,332,653	16,600,536
Hispanic	4,772,429	2,803,678	3,462,599	2,949,929	13,988,635	33,738,896
Color	12,612,146	7,067,641	8,591,788	7,529,264	35,800,840	71,591,664
White (%)	67.60	76.74	69.48	70.29	70.83	75.45
Black (%)	16.89	11.86	14.67	22.34	16.12	10.82
Asian (%)	6.37	4.32	6.28	2.26	5.08	4.71
Native American (%)	0.34	0.28	0.43	0.34	0.35	1.03
Other (%)	8.80	6.80	9.14	4.76	7.63	7.99
Hispanic (%)	14.92	11.62	15.54	16.69	14.56	16.23
Color (%)	39.43	29.29	38.57	42.60	37.26	34.44
Renter-occupied housing units (%)	36.28	26.38	29.32	28.18	30.69	28.73
College degree or higher (%)	31.44	30.66	30.49	26.07	30.03	26.9
Unemployed (%)	8.37	7.97	7.79	8.13	8.09	7.84
Below poverty line (%)	13.97	11.56	11.47	14.09	12.81	14.28
Mean household income ($)	77,805	77,145	78,260	67,522	75,845	68,593

Data source GeoLytics (2012c)

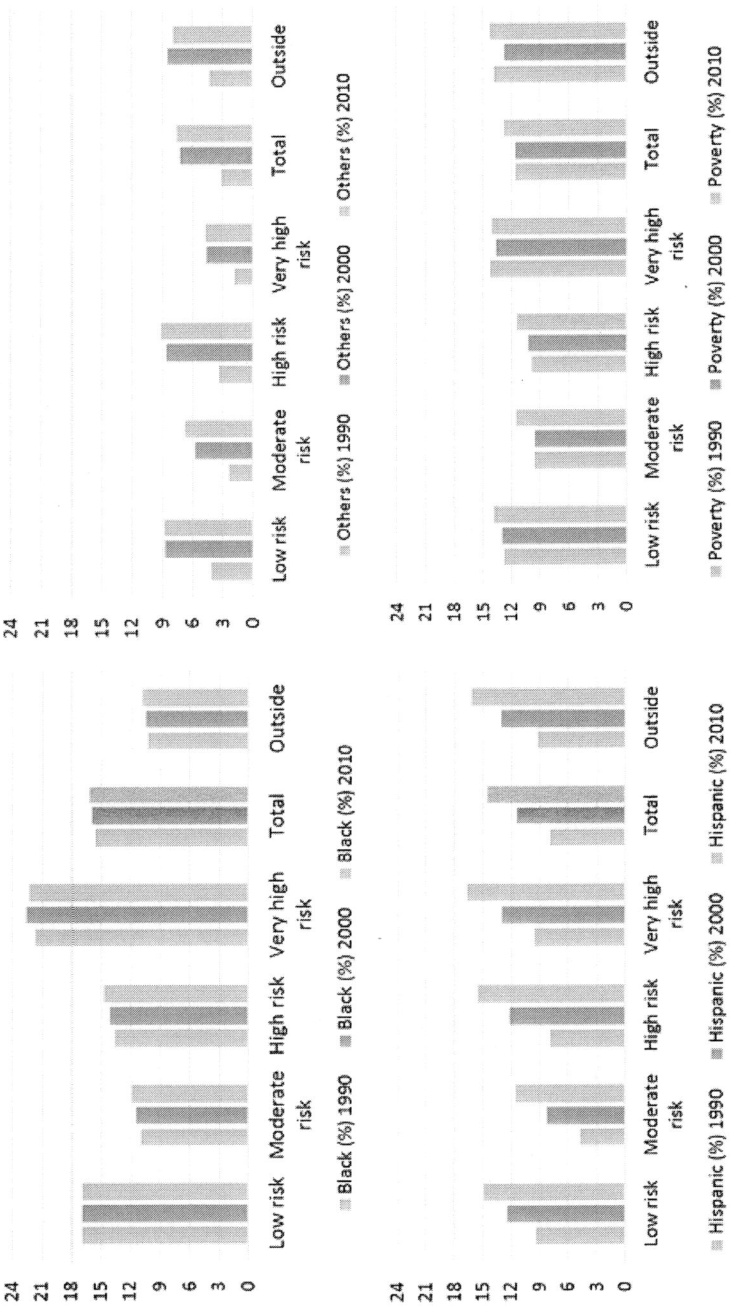

Figure 5. Demographic characteristics of populations living within a 50-mile radius of a NPP, as sorted by NPP PRI categories in 1990, 2000, and 2010

In addition to the highest percent Black (22.34 %) occurring among the host communities of NPPs classified as being at very high risk, these same NPPs were also associated with the highest percent Color (42.60 %) (Table 3). Similar patterns were observed in the data from 1990 to 2000 (Fig. 4).

In short, the minority groups who are in low socioeconomic status inevitably shoulder the burden of potential risks associated with the NPPs. The study provides two notions in regard to demographic characteristics of populations and their exposure to the four levels of risks. One notion is that there are variations in exposure to the four levels of risks among the populations living within a 50-mile radius. The other notion is that the minority groups living within a 50-mile radius inevitably expose to the four different levels of risks while their counterparts in the outlying areas do not expose to the risks. According to Bullard (1996, p. 493), "all people and communities are entitled to equal protection of environmental and public health laws and regulation." It is obvious that the communities living within a 50-mile radius from the nearest NPP faced environmental justice issues (Fig. 5). The findings reflect the fact that authorities made environmental decisions on siting of U.S. NPPs without incroporating consideration of unequal impacts from environmental hazards (Kyne 2015).

4.2 Demographic Trends by PRI Category

Changes over time in the demographics of the populations associated with each of the four PRI categories across two periods—1990–2000, and 2000–2010 were also examined. Between 2000 and 2010, percent Asian showed the greatest increase, up 48 % in areas classified as moderate risk; in those same areas, meanwhile, percent people living below the federal poverty line increased 1.97 % points. Percent Hispanic and percent Black registered the most pronounced increases, 44 and 17 %, respectively, in areas classified as high risk. Overall, percent Color showed the greatest increase in the high risk category, up 54 and 30 %, respectively, in 1990–2000 and 2000–2010.

The findings show that minority racial and ethnicity groups including Asian, Black, Color, Hispanic, and Native American increased during the two study periods, 1990–2000 and 2000–2010 (Table 4; Fig. 6).

Table 4. Changes in demographic characteristics of populations living within a 50-mile radius of a NPP, classified according to four PRI categories using 2010 U.S. Census data

Percent change	Low risk	Moderate RISK	High risk	Very high risk	Total	Outside	Year on year
Asian	59 %	71 %	62 %	45 %	61 %	40 %	1990–2000
Asian	35 %	48 %	46 %	44 %	41 %	38 %	2000–2010
Black	7 %	15 %	21 %	16 %	13 %	16 %	1990–2000
Black	2 %	10 %	17 %	6 %	7 %	13 %	2000–2010
Color	33 %	50 %	54 %	33 %	40 %	46 %	1990–2000
Color	12 %	28 %	30 %	17 %	20 %	25 %	2000–2010
Hispanic	42 %	92 %	80 %	52 %	59 %	62 %	1990–2000
Hispanic	22 %	49 %	44 %	37 %	35 %	36 %	2000–2010
Native American	28 %	20 %	33 %	31 %	28 %	20 %	1990–2000
Native American	−6 %	−11 %	−6 %	−4 %	−7 %	3 %	2000–2010
Other	120 %	161 %	193 %	174 %	150 %	128 %	1990–2000
Other	3 %	25 %	19 %	9 %	12 %	3 %	2000–2010
Below poverty line (%)	0.13	0.01	0.35	−0.67	−0.03	−1.10	1990–2000
Below poverty line (%)	1.00	1.97	1.25	0.45	1.17	1.55	2000–2010

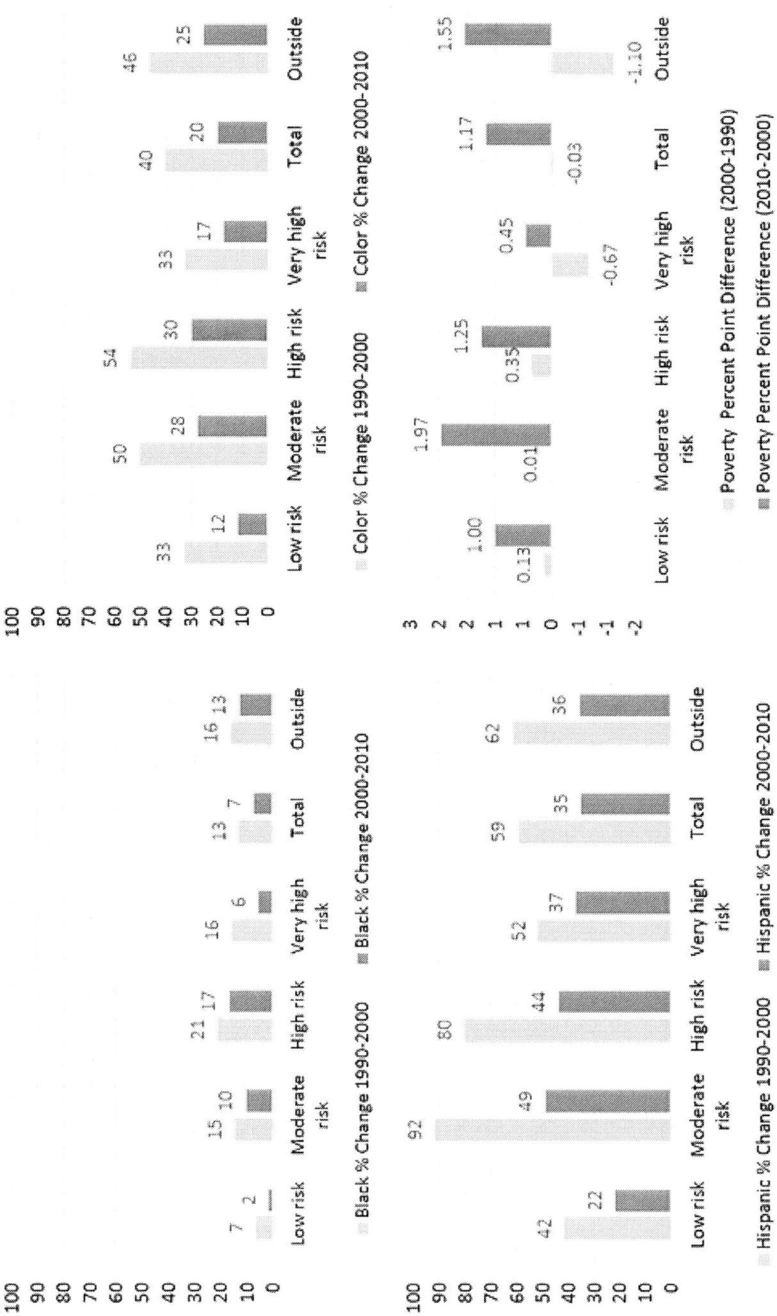

Figure 6. Changes in demographic characteristics of populations living in host communities of NPPs falling into different PRI levels

5. CONCLUSION

This analysis has provided one potential means of quantifying the level of potential risk associated with a given NPP and has investigated the disparities that exist in the sociodemographic characteristics of the populations that are exposed to different levels of risk. Much of the potential risk associated with NPPs stems from the fact that radiation is produced and may be released by these facilities. Exposure to radiation, regardless of the amount, increases the risk of disease, specifically cancer. A pattern was identified showing that populations living in areas associated with the greatest amount of potential risk tended to include a higher percentage of minorities than did the average host community regardless of PRI category as of 1990, 2000, and 2010. This implies that among the at-risk population, different groups of people face different levels of risk associated with or attributable to NPPs. This indicates that there are two layers of environmental justice issues. One layer of environmental justice issues involves those people living in host communities that are within a 50-mile radius of a NPP, who face environmental justice issues deriving from the potential risks associated with the NPP and is made evident when comparing these groups with their counterparts that live outside of those areas (Tables 2, 3). The second layer of environmental justice issues is that, among the populations of the host communities, certain groups—or entire populations at certain NPPs—are exposed to different levels of potential risk, classified here as low risk, moderate risk, high risk, and very high risk, depending on the unique circumstances and location of the individual NPP.

In addition, when the sociodemographic trends over that same time frame were considered, the investigators found that the largest percent changes had occurred in the percent Hispanic and percent Black in the high-risk category. As a percent, the host community population living in moderate risk areas increased during the study period 1990–2010. This implies that the environmental justice issues are on-going and have been in existence for at least three decades. The long-term presence of environmental justice issues facing the NPP host communities, coupled with the large size of such communities—estimated to be more than one third of the total U.S. population.

The study provided findings that there are environmental justice issues as a result of environmental decision making of siting NPPs without considering unequal impacts of hazards from them. The environmental justice issues have negative impacts on health and wellbeing of people living near the power plants (Kyne 2015). In fact, the environmental justice issues present enormous challenges to all stakeholders seeking to address these issues. In case of extreme events, the emergency evacuation will be very complex and will pose challenges to people who live near the NPPs. We think that participatory decision making among major stakeholders could not be avoided to collectively solve the environmental justice issues.

REFERENCES

1. Bullard, R.D. 1996. Environmental justice: It's more than waste facility siting. *Social Science Quarterly* 77(3): 493–499.

2. Denton, H.R. 1987. The causes and consequences of the Chernobyl nuclear accident and implications for the regulation of U.S. nuclear power plants. *Annals of Nuclear Energy* 14(6): 295–315.

3. Eisenhower, D. D. 1953. Atoms for peace. http://voicesofdemocracy.umd.edu/eisenhower-atoms-for-peace-speech-text/. Accessed 12 Aug 2012.

4. FEMA (Federal Emergency Management Agency). 2013. Earthquake hazard maps. http://www.fema.gov/earthquake/earthquake-hazard-maps. Accessed 14 Aug 2014.

5. Garrick, J. 2003. The current status, safety, and transportation of spent nuclear fuel. *The Bridge* 33(3): 11–18.

6. GeoLytics. 2012a. 2000 Census long form in 2010 boundaries. http://www.geolytics.com/USCensus,2000-Long-Form-in-2010-Boundaries,Products.asp. Accessed 14 Aug 2012.

7. GeoLytics. 2012b. Census 1990 in 2010 boundaries. http://www.geolytics.com/USCensus,Census-1990-in-2010-Boundaries,Products.asp. Accessed 14 Aug 2012.

8. GeoLytics. 2012c. Census 2010 / American community survey (ACS). http://www.geolytics.com/USCensus,Census2010ACS,Categories.asp. Accessed 14 Aug 2012.

9. Harris, J.T. 2013. US nuclear power plant radioactive effluents 1994–2009. Center for Advanced Energy Studies. Received the data via email, 20 June 2013.

10. Harris, J.T., and D.W. Miller. 2008. Radiological effluents released by US commercial nuclear power plants from 1995–2005. *Health Physics* 95(6): 734–743.

11. Holt, M., and A. Andrews. 2007. Nuclear power plants: Vulnerability to terrorist Attack. CSR report for Congress. https://fas.org/irp/crs/RS21131.pdf. Accessed 15 Jun 2012.

12. ICRP (International Commission on Radiological Protection). 2007. *The 2007 recommendations of the International Commission on Radiological Protection.* Publication 103. Ann. ICRP 37. http://www.icrp.org/docs/ICRP_Publication_103-Annals_of_the_ICRP_37%282-4%29-Free_extract.pdf. Accessed 21 Mar 2015.

13. Koch, W. 2011. Earthquake readiness of U.S. nuclear power plants is unclear. *USA Today.* 25 August 2011.http://usatoday30.usatoday.com/tech/science/environment/story/2011-08-25/Earthquake-readiness-of-US-nuclear-power-plants-is-unclear/50141704/1. Accessed 10 Sept 2012.

14. Kyne, D. 2015. Public exposure to U.S. commercial nuclear power plants induced disasters. *International Journal of Disaster Risk Science* 6(3): 238–249.

15. Lochbaum, D. 2011. *The NRC and nuclear power plant safety in 2010: A brighter spotlight needed.* Union of Concerned Scientists: Citizens and Scientists for Environmental Solutions. http://www.ucsusa.org/sites/default/files/legacy/assets/documents/nuclear_power/nrc-2010-full-report.pdf. Accessed 15 Aug 2015.

16. Lochbaum, D. 2012. *The NRC and nuclear power plant safety in 2011: Living on borrowed time.* Union of Concerned Scientists: Citizens and Scientists for Environmental Solutions. http://www.ucsusa.org/assets/documents/nuclear_power/NRC-nuclear-safety-2011-report.pdf. Accessed 15 Aug 2015.

17. Lochbaum, D. 2013. *The NRC and nuclear power plant safety in 2012: Tolerating the intolerable.* Union of Concerned Scientists: Citizens and Scientists for Environmental Solutions. http://www.ucsusa.org/sites/default/files/legacy/assets/documents/nuclear_power/NRC-nuclear-safety-2012-report.pdf. Accessed 15 Aug 2015.

18. Lochbaum, D. 2014. *The NRC and nuclear power plant safety in 2013: More Jekyll, less Hyde.* Union of Concerned Scientists: Citizens and Scientists for Environmental Solutions. http://www.ucsusa.org/sites/default/files/legacy/assets/documents/nuclear_power/NRC-Nuclear-Safety-Report-2013.pdf. Accessed 15 Aug 2015.

19. Lochbaum, D. 2015. *The NRC and nuclear power plant safety in 2014: Tarnished gold standard.* Union of Concerned Scientists: Citizens and Scientists for Environmental Solutions. http://www.ucsusa.org/sites/default/files/attach/2015/03/nrc-2014-full-report.pdf. Accessed 15 Aug 2015.

20. Morris-Suzuki, T., D. Boilley, D. McNeill, and A. Gundersen. 2012. *Lessons from Fukushima.* Greenpeace International. http://www.greenpeace.org/international/Global/international/publications/nuclear/2012/Fukushima/Lessons-from-Fukushima.pdf. Accessed 14 Aug 2012.

21. Nebraska Energy Quarterly. 1997. *State's first nuclear plant buried near Lincoln*. http://www.neo.ne.gov/winter97/win97_12.htm. Accessed 10 Oct 2012.

22. Nicoletti, G., S. Scarpetta, and O. Boylaud. 2000. *Summary indicators of product market regulation with an extension to employment protection legislation.* Economics department working papers No. 226, ECO/WKP(99)18. Organisation for Economic Co-operation and Development (OECD). http://www.oecd.org/eco/outlook/1880867.pdf. Accessed 19 Aug 2012.

23. OECD (Organisation for Economic Co-operation and Development). 2008. *Handbook on constructing composite indicators: Methodology and user guide*. http://www.oecd.org/std/42495745.pdf. Accessed 30 Aug 2012.

24. Perrow, C. 1999. Normal accidents: Living with high-risk technologies: With a new afterword and a postscript on the Y2K problem. New Jersey: Princeton University Press.

25. Richards, S., T. Frye, J. Shepherd, T. Nicholson, G. Kuzo, U. Shoop, S. Sakai, R. Allen, et al. 2011. *Liquid radioactive release lessons learned task force final report*. United States Nuclear Regulatory Commission (U.S. NRC). http://pbadupws.nrc.gov/docs/ML0626/ML062650312.pdf. Accessed 10 Aug 2012.

26. Tate, E., S.L. Cutter, and M. Berry. 2010. Integrated multihazard mapping. *Environment and Planning B:Planning and Design* 37(4): 646–663.

27. The Daily Beast. 2011. *Complete methodology: Most vulnerable power plants.* http://www.thedailybeast.com/articles/2011/03/16/most-vulnerable-power-plants-complete-methodology.html. Accessed 22 Aug 2012.

28. Three Mile Island Alert. 2013. *Incident chronology at TMI from NRC: 1979–2012*. http://www.tmia.com/node/1586. Accessed 17 Dec 2013.

29. United States Census Bureau. 2010. *TIGER/Line® Shapefiles pre-joined with demographic data: 2010 census summary file 1 demographic profile (DP1)*. ftp://ftp2.census.gov/geo/tiger/TIGER2010/TTRACT/2010/. Accessed 23 Aug 2012.

30. UNSCEAR (The United Nations Scientific Committee on the Effects of Atomic Radiation). 2012. *The Chernobyl accident: UNSCEAR's assessments of the radiation effects*. http://www.unscear.org/unscear/en/chernobyl.html. Accessed 15 Aug 2013.

31. U.S. GAO (U.S. Government Accountability Office). 2012. *Spent nuclear fuel: Accumulating quantities at commerical reactors present storage and other challenges* (GAO-12-797). http://www.gao.gov/assets/600/593745.pdf. Accessed 14 Aug 2012.

32. USGS (United States Geological Survey). 2013. *2008 United States national seismic hazard maps*. http://pubs.usgs.gov/fs/2008/3018/pdf/FS08-3018_508.pdf. Accessed 23 Jun 2014.

33. USGS (United States Geological Survey). 2014. *U.S. volcanoes and current activity alerts*. http://volcanoes.usgs.gov/. Accessed 14 Mar 2014.

34. U.S. NAS (United States National Academy of Sciences). 2006. *Safety and security of commercial spent nuclear fuel storage: Public report*. Washington, DC: Committee on the Safety and Security of Commercial Spent Nuclear Fuel Storage, National Research Council of the National Academies. http://www.nap.edu/catalog/11263/safety-and-security-of-commercial-spent-nuclear-fuel-storage-public. Accessed 20 Oct 2013.

35. U.S. NAS (United States National Academy of Sciences)/NRC (United States Nuclear Regulatory Commission). 2006. *Health risks from exposure to low levels of ionizing radiation: BEIR VII Phase 2*. Washington, DC: Board on Radiation Effects Research, National Research Council of the National Academies.http://www.nap.edu/catalog/11340/health-risks-from-exposure-to-low-levels-of-ionizing-radiation. Accessed 15 Aug 2015.

36. U.S. NRC (United States Nuclear Regulatory Commission). 2011. *Nuclear Regulartory Commission*. http://www.nrc.gov. Accessed 14 Aug 2012.

37. U.S. NRC (United States Nuclear Regulatory Commission). 2012a. *1Q/2012 ROP performance indicators summary*. http://www.nrc.gov/NRR/OVERSIGHT/ASSESS/2012q1/pi_summary.pdf. Accessed 14 Aug 2012.

38. U.S. NRC (United States Nuclear Regulatory Commission). 2012b. *Information digest, 2011–2012* (NUREG-1350, Volume 23). http://pbadupws.nrc.gov/docs/ML1124/ML11241A096.pdf. Accessed 19 Aug 2012.

39. U.S. NRC (United States Nuclear Regulatory Commission). 2013a. *Backgrounder on chernobyl nuclear power plant accident*. http://www.nrc.gov/reading-rm/doc-collections/fact-sheets/chernobyl-bg.html. Accessed 19 Aug 2013.

40. U.S. NRC (United States Nuclear Regulatory Commission). 2013b. *Transportation of spent fuel and radioactive materials*. http://www.nrc.gov/reading-rm/doc-collections/fact-sheets/transport-spenfuel-radiomats.pdf. Accessed 19 Nov 2013.

41. U.S. NRC (United States Nuclear Regulatory Commission). 2013c. *Backgrounder on decommissioning nuclear power plants*. http://www.nrc.gov/reading-rm/doc-collections/fact-sheets/decommissioning.html. Accessed 17 Aug 2013.

42. U.S. NRC (United States Nuclear Regulatory Commission). 2013d. *Fact sheet on summary of Japan events in March 2011 and NRC response*. http://www.nrc.gov/reading-rm/doc-collections/fact-sheets/fs-japan-events.html. Accessed 15 Aug 2013.

43. U.S. NRC (United States Nuclear Regulatory Commission). 2013e. *Spent fuel transportation: Risk assessment*(NUREG-2125). Office of Nuclear Material Safety and Safeguards, United States Nuclear Regulatory Commission. http://pbadupws.nrc.gov/docs/ML1212/ML12125A218.pdf. Accessed 24 Aug 2013.

44. U.S. NRC (United States Nuclear Regulatory Commission). 2014. *Status of the decommissioning program.*http://pbadupws.nrc.gov/docs/ML1429/ML14294A239.pdf. Accessed 24 Mar 2014

45. U.S. NRC (United States Nuclear Regulatory Commission). 2015. *Information digest, 2015–2016*. http://www.nrc.gov/reading-rm/doc-collections/nuregs/staff/sr1350/v26/sr1350v26.pdf. Accessed 23 May 2015.

46. Wald, M. L. 2009. Future dim for nuclear waste repository. *The New York Times*. 5 March 2009. http://www.nytimes.com/2009/03/06/science/earth/06yucca.html?ref=yuccamountain&pagewanted=print. Accessed 10 Aug 2012.

47. World Nuclear Association. 2013. *Safety of nuclear power reactors*. http://www.world-nuclear.org/info/Safety-and-Security/Safety-of-Plants/Safety-of-Nuclear-Power-Reactors/. Accessed 17 Aug 2013.

48. Zeller, T. 2011. Experts had long criticized potential weakness in design of stricken reactor. *New York Times*. 15 March 2011. http://www.nytimes.com/2011/03/16/world/asia/16contain.html?_r=0&pagewanted=print. Accessed 14 Aug 2012.

Chapter 11

On the Burning of Plutonium Originating from Light Water Reactor Use in a Fast Molten Salt Reactor—A Neutron Physical Study

Bruno Merk [1,2,*] **and Dzianis Litskevich** [1]

[1]Helmholtz-Zentrum Dresden-Rossendorf, Bautzner Landstrasse 400, Dresden 01328, Germany

[2]Center for Materials and Structure, University of Liverpool, Liverpool, Merseyside L69 3BX, UK

ABSTRACT

An efficient burning of the plutonium produced during light water reactor (LWR) operation has the potential to significantly improve the sustainability indices of LWR operations. The work offers a comparison of the efficiency of Pu burning in different reactor configurations—a molten salt fast reactor, a LWR with mixed oxide (MOX) fuel, and a sodium cooled fast reactor. The calculations are performed using the HELIOS 2 code. All results are evaluated against the plutonium burning efficiency determined in the Consommation Accrue de Plutonium dans les Réacteurs à Neutrons RApides (CAPRA) project. The results are discussed with special view on the increased sustainability of LWR use in the case of successful avoidance of an accumulation of Pu which otherwise would have to be forwarded to a final disposal. A strategic discussion is given about the unavoidable plutonium production, the possibility to burn the plutonium to avoid a burden for the future generations which would have to be controlled.

KEYWORDS

plutonium; plutonium management; plutonium burning; nuclear reactors; molten salt reactor (MSR); fast reactor; light water reactor; mixed oxide (MOX) fuel

1. INTRODUCTION

1.1. The Plutonium Problem

Today, almost all kinds of nuclear reactors are operated using ^{238}U-containing nuclear fuel. Due to the unavoidable breeding processes higher isotopes are formed besides the production of energy and the formation of the fission products. Consequently, the operation of almost all kinds of nuclear reactors leads to the built-up of artificial isotopes like plutonium as well as other higher transuranium (TRU) isotopes. Most of the created artificial isotopes are characterized by very long radioactive half-lives, which have to be considered when a final disposal for the nuclear waste is designed.

"One of the greatest challenges in the use of nuclear energy is the highly radioactive waste which is generated during power production. It must be dealt with safely and effectively. While technical solutions exist, including deep geological repositories, progress in the disposal of radioactive waste has been influenced, and in many cases delayed, by public perceptions about the safety of the technology. One of the primary reasons for this is the long life of many of the radioisotopes generated from fission, with half-lives on the order of 100,000 to a million years. Problems of perception could be reduced to an essential degree if there were a way to burn or destroy the most toxic long-lived radioactive wastes during the production of energy" [1]. These are the words of Victor Arkhipov a consultant in the IAEA division of nuclear power and the fuel cycle, in the nuclear power technology development section given in an IAEA Bulletin in 1997. Following this arguments a very special focus should be given to the plutonium isotopes. On the one hand, these isotopes are by far the major share (more than 80%) of the TRU isotopes. On the other hand, plutonium has the potential for the production of nuclear weapons. "A number of issues arise because of plutonium's potential use as an energy source and for the production of nuclear weapons. ... Within 200 years, the protection afforded by intense radioactivity (spent fuel standard) will disappear as the result of the decay of most radioactive nuclides. If the spent fuel is buried in a geological formation, it might be regarded as a potential 'plutonium mine' meaning that at some later point in time the buried plutonium could be mined and extracted. ... Without such reactors (fast reactors), spent MOX fuel will still end up in a final depository or in storage facilities. ... Are there any other methods for burning excess plutonium?" [2]. These sentences are taken from a contribution of Noboru Oi, a senior staff member of the IAEA Department of Nuclear Energy, published in the IAEA Bulletin on Comparing Energy Options in 1998. Already some years before a discussion has been started on the political level by the RAND Corporation (Santa Monica, CA, USA). In a publication on the accumulation of weapon-usable plutonium the problems arising from non-separated and separated Pu from nuclear reactors is discussed, too. "The concern of the former (non-separated Pu) is that, eventually, the radiation from spent fuel will drop so much that national and subnational groups can reprocess the spent fuel quickly and easily for plutonium ... separated Pu is a more immediate and urgent issue, because the most difficult task of extracting Pu from the intensively radioactive spent fuel has already been performed; the remaining steps for incorporating the material into a nuclear bomb are much easier." [3]. The drop of the radiation level often call the decay of the self-protection level of Pu in the fuel is mostly dominated by the radiation level of the dominating fission products with intermediate half live time ~30 years. Consequently the radiation level drops to: ~10% after 100 years and ~1% after 200 years.

Recently, comparable concerns regarding the misuse of plutonium have been raised in the German P&T study in the discussion of the chances of P&T. "The application of P&T on industrial level has the potential for a plutonium content in the final disposal which is negligible after the application of P&T. Thus there is no risk of misuse and theft of plutonium (from the final disposal) anymore" [4,5].

Already in 1996 a NATO advanced research workshop has been held in Moscow to discuss the problem of the plutonium management on a scientific level in an international group of experts. In the introduction given by Merz the risks arising from plutonium has been highlighted: "The lower content of the plutonium isotope ^{239}Pu in energy-grade plutonium, as well as the presence of a comparatively large amount of isotope ^{240}Pu, make energy grade plutonium a less preferably material for nuclear munitions production than weapon-grade. Nevertheless, energy grade plutonium can be directly used for making a nuclear explosive device. All things being equal, the device will only be larger if it was made of weapons-grade plutonium" [6].

Generally, it can be concluded based on the different citations given above that a method for burning of the already existing and the in future in light water reactors (LWRs) produced plutonium is of high interest. The efficient and complete burning of the plutonium produced during LWR operation offers the potential to increase the sustainability of the LWR operation significantly. This can be stated at least for all countries which do not plan to switch their nuclear energy production capacities from the current park of LWRs to a significant park of fast reactors applying a closed fuel cycle, except from India which plans to build a significant number of fast reactors [7], no country envisages this large park of fast reactors currently. Only the efficient and root-and-branch burning of plutonium is the ultimate argument to avoid the risk misuse of plutonium in perpetuity.

One of the first promising coordinated scientific project on the burning of TRU isotopes and plutonium, especially, has been started in the early 90s in France—the Consommation Accrue de Plutonium dans les Réacteurs à Neutrons Rapides/Consommation Accrue des Dechets dans les Reacteurs RApides (CAPRA/CADRA) project, CAPRA—how to burn Plutonium, CADRA how to burn minor actinides. A short overview of the CAPRA project has been given by Languille *et al.*[8]: "During the first two-year phase (93–94) of the CAPRA Project studies, the feasibility of a Fast Reactor (FR) plutonium burner which also contributes to the destruction of Minor Actinides (MA) has been assessed for the oxide reference option based on a fuel plutonium content of about 45%. The evaluation of this core of 1500 MWe was carried out covering different aspects: fuel behaviour, physics, and safety. Based on this evaluation, the feasibility of such a burner core is demonstrated. The heterogeneous concept of dilution provides a large flexibility to the core but a low fuel residence time".

Thus, the current level the fast reactor technology does not allow implementing of the fast breeder reactors for industrial scales of burning of plutonium. Therefore other optimized opportunities, such as molten salt fast reactors, can be taken into consideration.

1.2. Why Molten Salt Reactors (MSRs)

Already in 1992, in a technical note in the ANS journal Nuclear Technology based on their experience, Gat *et al.* [9] proposed the use of "Molten Salt Reactors (MSRs) for Burning Dismantled Weapons Fuel". In their abstract they summarized some really strong arguments which make MSRs highly interesting for the burning of plutonium: "The Molten-salt reactors have the flexibility to utilize any fissile fuel in continuous operation with no special modifications, as demonstrated in the Molten Salt Reactor Experiment (MSRE), while maintaining their economy. The MSRs further require a minimum of special fuel preparation and can tolerate denaturing and dilution of the fuel. Fuel shipments can be arbitrarily small, which may reduce the risk of diversion. The MSRs have inherent safety features that make them acceptable and attractive. They can burn a fuel type completely and convert it to other fuels. The MSRs also have the potential for burning the actinides and delivering the waste in an optimal form, thus contributing to the solution of one of the major remaining problems for deployment of nuclear power". The inherent safety features are based on the strong feedback effects in molten salt fast reactors (MSFRs).

The major effect, the salt density effect is strongly negative in all cases, due to the unity of coolant and fuel in a MSR. A decrease in the salt density due to a temperature increase coincides with a decrease of the density of the fissile material dissolved in the salt in the control volume. Thus the MSFR is a very stable system with strong feedback effects which inherently tend to stabilize the reactor at the given power level. The flexibility of the reactor for the use of different fissile materials has been demonstrated by the smooth transition from the ^{235}U startup core of the MSRE to the core fully operated on ^{233}U in the first months of 1969. This was the first time ^{233}U had been used as a reactor fuel [10].

Already in 1978 Engel *et al.* [11], highlighted one of the most attractive features of liquid fueled reactors. When the plutonium or other fissile material is once put into a molten salt reactor system it is not necessary to separate the material in any of the steps of the burning operation anymore. This is a strong advantage which is in strong contrast to the fuel management like it is to be applied for any kind of solid fueled reactor with the separated steps of reprocessing, fuel production, and fuel irradiation. Unfortunately, in most of the cases, the different steps can even not be conducted at one site. Thus there are transports required between the sites. In a follow up publication, the authors discussed the reduction of the proliferation risk as one important aspect in the burning of plutonium to achieve a reduction of the plutonium amounts in the nuclear waste of LWRs. "MSRs, because of the fluid nature of the fuel, appear to provide an attractive approach to efficient fuel utilization in the thorium-^{233}U cycle as well as a means for limiting the availability of plutonium and the general proliferation risks associated with nuclear power generation. ... the plutonium (most of which is consumed *in situ*) would be of poor quality and would never be isolated from all other undesirable nuclides. Thus, such systems would provide for efficient utilization of uranium resources in a proliferation-resistant environment while limiting the amount of plutonium (and transplutonium actinides) that would have to be handled as waste" [12].

A very detailed analysis and comparison of the general features of sodium cooled fast reactors and MSRs with the special view on the transmutation of TRU isotopes has been given at the international conference on fast reactors and related fuel cycles: safe technologies and sustainable scenarios (FR13),

Paris [13] and later extended [14]. In these publications clear operational as well as safety and efficiency advantages of MSRs are carved out.

A presentation on the possible application of MSRs for burning of plutonium has already given at the mentioned NATO workshop in Moscow. The authors highlighted the possibility of an "all in one" solution and the flexibility of the system for the use of different fuels, too. "The advantages of the MSR as a burner reactor follow not only form a possibility for its combination with the gas-fluoride technique of fuel reprocessing, which is low-cost and produces a small quantity of waste, but also from its capability to use fuel of any nuclide composition without Pu purification from MA and FP. The MSRs have the flexibility to utilize any fissile fuel in continuous operation with no special modification of the core as demonstrated during MSRE operation for U3, U5 and Pu" [15].

From the point of view of the complete nuclear fuel cycle, MSRs have been a part of the U.S. study on Comprehensive fuel cycle options. In the study some really interesting and important effects have been carved out in the comparison with different other kinds of reactors. "... when a MSR is adopted as the AE (Analysis Example) of a specific EG (Evaluation Groups), the material loss fraction was assumed to be lower than that of the common assumption because the molten-salt fuel is continuously reprocessed in an online mode. ... Even though uniform assumptions for losses were used, the exception of having to specify losses for an MSR differently (0.2% rather than 1.0%) caused the AE of EG26 to have the highest (fissile material) utilization because the exception resulted in the lowest material losses being assumed for the MSR that was utilized as the AE of EG26" [16]. These assumptions result in the best natural resource utilization in the whole study. Additional facts to be highlighted are: No spent nuclear fuel is created in MSRs for EG26 in the analysis and no depleted uranium, no non-used recovered uranium, and no recovered thorium mass will be created; only high level waste in the form of fission products is left over. This leads to the lowest over all waste mass for MSRs in the analysis of the Comprehensive Fuel Cycle Options launched in the U.S. in 2010 [16].

Summarizing the citations given above, the following conclusions can be made concerning the advantages of the molten salt fast reactor technology:

- flexibility to utilize any fissile fuel in continuous operation
- minimum of special fuel preparation
- arbitrarily small fuel shipments
- inherent safety features

- limitation of the plutonium availability and reduction of the risk of proliferation
- possibility of combination with the gas-fluoride technique of fuel reprocessing
- possibility to use fuel of any nuclide composition without plutonium purification from minor actinides and fissile products
- fissile material which is once inserted is not needed to be taken out anymore.

All these excerpts together demonstrate the high attractiveness of the MSR technology for the application as plutonium or TRU burner for the major objective of an economic and efficient burning process. Exactly these objectives are the key points for the already mentioned required increase in the sustainability of the LWR operation achieved by an efficient and complete burning of the plutonium produced during LWR operation.

1.3. Development of MSRs

MSRs have a long history, based on a, somewhat curious idea, developed in the early phase of the nuclear development. In the late 40s and early 50s the project of the nuclear aircraft has been investigated. This idea led to the Aircraft Reactor Experiment, a small thermal reactor (2.5 MW) [10]. The first small scale experiment has been followed by a larger one, the MSRE with 8 MW thermal. "Design of the MSRE started in the summer of 1960 and construction started 18 month later, at the beginning of 1962. The reactor went critical in June 1965, and was briefly at full power a year later" [10].

Recently, the MSR technology has attracted some new interest. Many countries worldwide contribute to MSR technology, among which are the United States, the Russian Federation, the European Union, France, Japan, India and for the past few years China and Korea. In Europe interest has been focused in the EURATOM project MOST—review on MSR Technology [17,18]. Following the MOST project, in 2010, two projects on MSR were selected and funded by European and Russian Agencies: the European Project: "Evaluation and Viability of Liquid fuel fast reactor systems" (EVOL) [19,20,21] and the Russian Project: "Minor Actinides Recycling in molten salts" (MARS) [22,23,24]. A three-year Euratom-Rosatom collaboration has been established, through the parallel coordinated projects (MARS-EVOL) on MSR [25]. In contrast to the historic MSRE, the objective of the new projects is the development of a molten salt reactor with fast neutron spectrum. This is achieved by omitting the graphite which has been used for moderation in the MSRE. The major argument, for the change in the neutron spectrum, is the improved sustainability indices. The MSRs are planned to be suitable for power production and either for breeding or for the transmutation of actinides from spent nuclear fuel.

2. MATERIALS AND METHODS [26]

2.1. Reference Configurations

2.1.1. Molten Salt Fast Reactor (MSFR)

The calculation for the MSFR is based on the core dimensions and the boundary conditions given in the EVOL benchmark definition (see Figure 1). The reference core is a MSFR with 3000 MW_{th} and a fast neutron spectrum. In this first approach the core is a single cylinder. The nuclear reactions occur within the flowing fuel salt inside the cylinder [27]. The dimensions are shown in Figure 1. The core is composed of four volumes: the active core, the upper extraction area, the lower injection area, and the out of core area with the heat exchanger and the pumps.

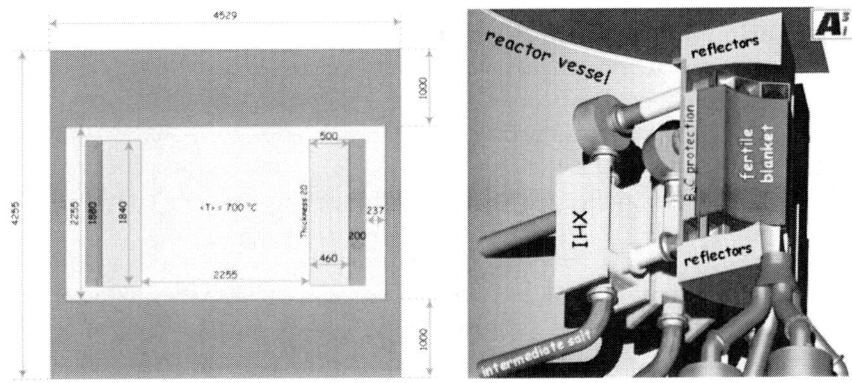

(a) (b)

Figure 1. (**a**) Simplified scheme of the circular molten salt fast reactor (MSFR) system (side view) including the core, blanket and intermediate heat exchangers (IHX); (**b**) benchmark definition [27].

For the simulations, the HELIOS 2.1 code is used [28]. The benchmark configuration is transferred to a volume corrected 2D circular HELIOS model (see Figure 2) which is adapted to 3D using a buckling correction available in HELIOS (BSQ: 0.00002). The value has been fixed by a comparison of 2D and 3D calculations within the EVOL benchmark exercises. The leakage in radial direction is directly modelled. Additionally, the model has been detailed compared to the EVOL benchmark configuration. The 16 heat exchanger pipes are resolved for a better representation of the real geometry.

The HELIOS code package is an industrial standard software which performs the neutron transport calculation, the burn up calculation, and if requested the cross section preparation in defined calculation areas. The code has been written for the simulation of solid structured fuel assemblies. Thus, the possibility of online refueling and online reprocessing is not foreseen. To deal with this special feature a PYTHON script has been developed, see Figure 3. All input data, which is constant during the whole reactor operation, is stored in an expert input. The changing material configuration is given in the user input which is written new in every cycle using the PYTHON script. Within each cycle five burnup steps of 1000 MWd/tHM at a power of 143 W/g are calculated in one HELIOS run. Both inputs are merged in the pre-processor AURORA. The results are finally evaluated for each cycle in the post-processor ZENITH. Here it is decided which isotopes will be fed back into the next user input. Theoretically, this scheme could simulate the feeding process accurately by using small time steps. Two different time scales are modelled for the salt cleanup. These are the helium bubbling for the gaseous and the volatile fission products (these fission products are not considered when a new cycle is started, thus all fission products are taken out of the system after 5000 MWd/tHM) and the online salt cleanup for the dissolved fission products (only 16.6% of the general soluble fission products and 15% the lanthanides are taken out at the end of the cycle, the rest is carried forward via the script to model the permanent presence of a part of the fission products). A more detailed description of the modelling is given in [29].

One of the salt configurations and the corresponding thermodynamic data of the molten salt actinide recycler and transmuter (MOSART) project [23] is used, 17LiF-58NaF-25BeF$_2$ (mole%) mixed with TRUF$_3$, which has been developed for the use in a fertile free system. The salt composition in the fertile region is pure LiF.

2.1.2. Sodium Cooled Fast Reactor (SFR) Fuel Assembly

The calculations for the fast reactor configuration are carried out only on fuel assembly level like it is standard in solid fueled reactor systems for the preparation of the cross sections for the following full core calculations. The fast reactor fuel assembly model is derived from the data for the European Fast Reactor. The data is mostly given in the IAEA Fast Reactor Database—2006 Update [30]. Additional data has been taken from the textbook Waltar, Reynolds: Fast Breeder Reactors [31] and from the published European Fast Reactor (EFR) fuel assembly design [32], see Table 1.

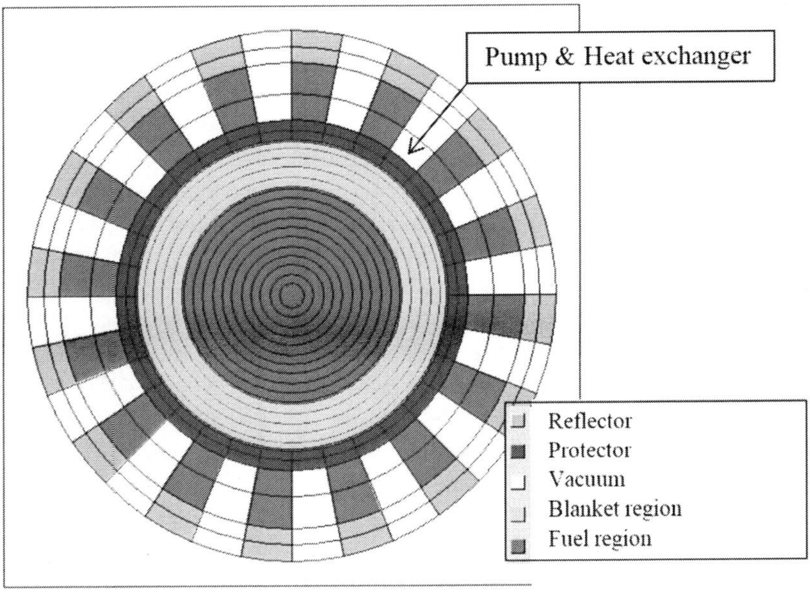

Figure 2. Volume corrected 2D circular HELIOS model of the molten salt reactor (MSR) (top view of the reactor).

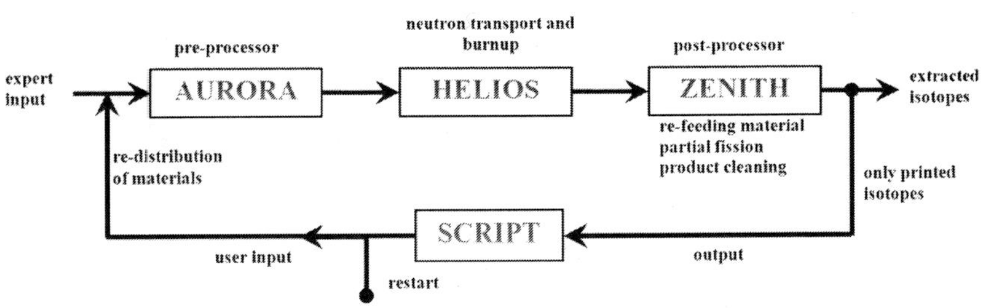

Figure 3. Description of the calculation cycle for the simulation of a MSR.

Cladding, wrapper wire and can wall are made from stainless steel, SS 304. The used material configuration definition is following the HELIOS definition [28]. The temperatures are: 635 °C for the cladding, 545 °C for the wrapper and the can wall, and 1100 °C for the fuel.

The modeled geometric arrangement of the reference system with 10 rings of fuel rods following the EFR design is shown in Figure 4 for 1/6 part of one fuel assembly. The power density is set to 118.8 W/g corresponding to the maximum power density in the EFR. The fuel assemblies contain 18.3%, 22.4%, and 26.9% plutonium (depending on the core zone) and a burnup of 200 GWd/tHM is used for the comparison. The value 200 GWd/tHM has been chosen based on data from [8,33] as a typical for this type of reactor.

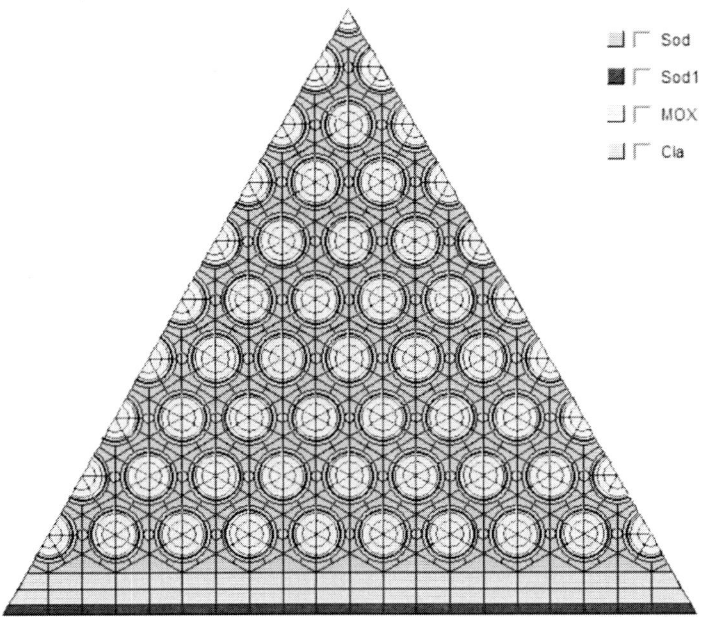

Figure 4. 1/6 of the sodium cooled fast reactor fuel assembly corresponding to the European fast reactor (EFR) geometry.

2.1.3. Light Water Reactor Mixed Oxide (LWR MOX) Fuel Assembly

The calculations for the light water reactor configuration with MOX fuel are carried out only on fuel assembly level like it is standard in solid fueled reactor systems for the preparation of the cross sections for the following full core calculations. The fuel assembly model is following the OECD/NEA AND U.S. NRC PWR MOX/UO2 CORE TRANSIENT BENCHMARK [34] definition. The MOX fuel assembly consists of three different types of pins using 2.5% (MOX 1), 3% (MOX 2) and 5% Pu fiss (MOX 3) content, and WABA—wet annular burnable absorber in the guide tubes.

Table 1. Basic geometric data of the fast reactor fuel assembly.

Component	Dimension	Unit
Outer pin diameter	8.5	mm
Cladding thickness	0.52	mm
Pitch to diameter	1.2	-
Can wall thickness	4.5	mm
Fuel density	9.26	g/cm^3

The fuel assembly consists of a 17 by 17 arrangement with 264 fuel pins (see Table 2), 24 guide tubes, and one instrumentation tube. The geometric arrangement of the HELIOS model is given in Figure 5. The average Pu_{fiss} content in the fuel assembly is calculated to 4.3% Pu_{fiss} using for this study reactor grade plutonium and a fuel assembly burnup of 50 GWd/tHM at a power of 37.11 W/g.

Table 2. Basic geometric data of the light water reactor mixed oxide (LWR MOX) fuel assembly.

Component	Dimension	Unit
Outer pin diameter	9.166	mm
Cladding thickness	0.573	mm
Pin pitch	12.6	mm
Assembly pitch	21.42	cm
Fuel density	10.41	g/cm^3

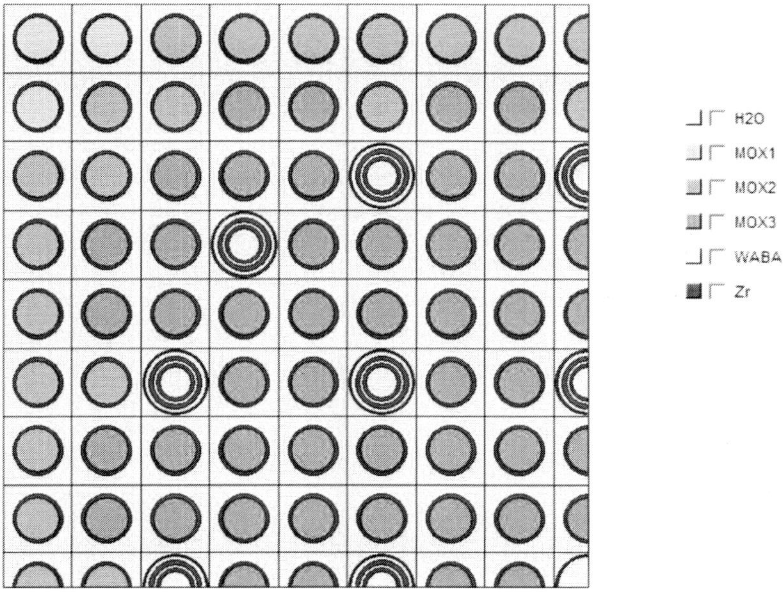

Figure 5. 1/4 of the light water reactor mixed oxide (LWR MOX) fuel assembly following the OECD/NEA AND U.S. NRC PWR MOX/UO2 CORE TRANSIENT BENCHMARK geometry.

2.2. Calculation Tool and Material

For all comparison calculations, the HELIOS 2.1 code has been used. The code is a 2D spectral code with wide unstructured mesh capabilities and a transport solver, based on the collision probability method [35]. The used libraries are the code internal 177 energy group library for the LWR and Sodium Cooled Fast Reactor (SFR) calculations and the 47 energy group library [28] for the MSR calculation (due to the significantly higher number of required calculations). The library id based on ENDF/B VII. Due to the characteristics of HELIOS, some approximations have to be accepted for the MSR and the fast reactor application. There is no fuel salt movement, thus an undesired burnup distribution arises during one calculation cycle. The materials are only re-distributed when a new user input is defined by the PYTHON script.

The HELIOS package has been developed as LWR code and a LWR spectrum is used for the weighting of the master libraries inside each energy group. A significant improvement has been observed for the use of HELIOS 2 with the cross section set based on ENDF/B VII. Comparisons to SERPENT on the isotope accumulation during the burnup in a fast reactor configuration using different

HELIOS libraries have shown good agreement for the major isotopes [36]. Additionally, different older versions of the HELIOS 1 code have been used in the EVOL benchmark calculations for MSFR [37] and in the benchmark calculations in the ESNII+ project for SFR [38]. In both benchmark calculations the results were on a comparable level to the other applied codes. No major systematic discrepancies have been observed. The approximations and the use of the HELIOS code package seem to be adequate for this kind of cross over investigation for different reactors with significantly different neutron spectra.

The used TRU isotopic vector (see Table 3) is configured in the case of burning of 4% enriched uranium oxide (UOX) fuel in a LWR with 50 GWd/tHM burnup. The fuel cycle times are 5 years storage before reprocessing and 2 years until reuse in the reactor.

Table 3. Pu vector from 4% enriched uranium oxide (UOX) fuel with 50 GWd/tHM burnup. 5 years storage before reprocessing and 2 years until reuse in the reactor.

Isotope	Content	Unit
Pu-238	2.62	%
Pu-239	54.43	%
Pu-240	23.70	%
Pu-241	11.28	%
Pu-242	6.83	%
Am-241	1.14	%

2.3. Results and Discussion

The results presented in Table 4 demonstrate efficiency of the molten salt fast reactors as transmuter in comparison with a sodium cooled fast reactor and a thermal light water reactor with MOX fuel. The Pu reduction rate depends here only on the operation time and the remaining rest of fuel at the end of the operation time. The rest after 60 and 80 years is almost identical. However, this remaining fuel salt with the rest of the plutonium could be either used in a follow up machine, or a twofold lifecycle could be applied with a deep burn phase in the last years of operation [13,14,29] to reduce the TRU amount in the end of life (EOL) core. The results for the LWR and for the SFR demonstrate the limits given by the solid fuel. In current types of reactors this solid fuel cannot be produced in a fertile free configuration. Consequently, the MOX fuel used in these reactors leads to the breeding of new, in the view of Pu burning, undesired plutonium. The energy produced during the burnup cycle using the inserted plutonium is in all cases in the same range, but the plutonium burning is comparably low. This can be observed for the LWR where the Pu amount is reduced by only 36%. The Pu burning in MOX fuels in LWR is currently only operated in a single step. Generally, a second step could be possible, but no real closed fuel cycle operation with multiple recycling. The reason for this limitation is given by the degradation of the Pu vector during Pu burning in a LWR spectrum, se the discussion of the EOL Pu vector below.

In the studied SFR configuration, the Pu reduction is even worse. This should not wonder at all, since the EFR has not been designed to burn Pu, rather the design target was breeding of fresh fissile material. Depending on the initial Pu content the Pu reduction varies. The inner fuel with the low Pu content leads only to a minor reduction since the breeding of fresh Pu almost compensates the burning. The increase of the initial Pu content improves the Pu reduction slowly—the configuration comes closer to a burner configuration. The results for the lattice calculation with the 26.9% initial Pu content should be taken with care due to the very high k_{inf}. In the real reactor configuration, these fuel assemblies act as a strong neutron source for their surrounding assemblies leading to breeding there. The initial Pu content is comparable to the fuel assemblies used in CAPRA, but there ~1/3 of the pins has been replaced by diluents to reduce the criticality. In the EFR full core configuration the breeding gain has been planned to −0.2 in the core only and +0.02 in the whole system including the blanket assemblies [30]. In general, it has to be kept in mind, that the typical SFR using solid fuel is foreseen to be operated in a closed fuel cycle mode. The flexibility of SFRs opens a whole range of operational modes-breeder, iso-breeder or burner, just depending on the core design [39]. However, the efficiency even of the burner configuration is limited since for the closed fuel cycle operation an acceptable Pu vector has to be preserved. This is usually achieved by breeding of fresh Pu. If this is not the case, the closed fuel cycle operation will not be possible as discovered during the CAPRA studies by Wiese, see discussion below. However, the closed fuel cycle which is foreseen to be established in the SFR operation has some important drawback on the time schedules. The fuel requires significant time to cool down before the reprocessing can be performed, at least as long as the reprocessing is performed in the traditional aqueous scheme. The reprocessing has to be followed by a new production of solid fuel which than can be inserted into the reactor. For the whole process a time period of 5 to 10 years out of the reactor has to be estimated for. Consequently, the fuel has to be handled out of the reactor several times in the SFR fuel cycle. This is in strong contrast to the MSR where the fuel is kept inside the system until it is burnt.

Table 4. Pu vector at end of life (EOL) and Pu burning rate for different reactor types. Molten salt fast reactor: MSFR; light water reactor mixed oxide: LWR MOX; sodium Cooled Fast Reactor: SFR.

Reactor type	MSFR		LWR MOX assembly	SFR assemblies		
Operational parameter	60 years	80 years	4.3% Pu_{fiss}	18.3%	22.4%	26.9%
Pu-238	3.9%	3.9%	0.7%	1.4%	1.7%	1.9%
Pu-239	22.4%	22.7%	45.5%	54.8%	52.1%	49.8%
Pu-240	46.6%	46.5%	30.1%	31.7%	32.6%	33.4%
Pu-241	10.0%	9.9%	17.6%	5.5%	6.0%	6.3%
Pu-242	17.0%	16.9%	6.1%	6.6%	7.6%	8.3%
Pu burning rate relative	83.9%	87.2%	35.9%	11.9%	21.8%	28.1%
Fuel target burnup	-	-	50 GWd/tHM	200 GWd/tHM	200 GWd/tHM	200 GWd/tHM
Inserted Pu share	100%	100%	6.54%	18.3%	22.4%	26.9%
Energy out of inserted plutonium	821 GWd/t	846 GWd/t	764 GWd/t	1093 GWd/t	893 GWd/t	743 GWd/t

The extremely high reductions of the Pu amount in the MSFR cases are a result of the fertile free configuration. Thus, energy is only produced from plutonium and no new plutonium is produced via breeding. The longer the machine operates, the better the Pu reduction is since the unloading mass is in all cases almost identical—the core is after 60 years of operation almost in equilibrium. In contrast the inserted mass increases slightly with every operation year due to the small built up of higher TRUs. The required solution capacity of the salt for the required TRU load can be assured over the whole operational span for the used salt defined in the MOSART project.

However, if breeding is welcome in the fuel cycle strategy, it would be possible in MSFR, too due the already mentioned high flexibility of MSRs. The big advantage of the MSFR in this case would be the elimination of the multi recycling and the solid fuel production. Additionally, no Pu, neither the inserted one, nor the bred one will come out of the reactor. Remember the statement: When the plutonium or other fissile material is once put into a MSR system it is not necessary to separate the material in any of the steps of the burning operation anymore.

The plutonium vector at EOL is an important indicator for the evaluation of the nuclear security. The higher the share of ^{239}Pu in the fuel at EOL is, the higher the proliferation risk is. The highest ^{239}Pu share at EOL is available in the SFR, since the breeding of new ^{239}Pu takes place most efficiently. The breeding process is slowed down with increasing initial Pu content. This fact has been applied and optimized in the CAPRA project by the design of fuel assemblies with very high initial Pu content and distributed so-called diluents made of steel [8]. In LWRs, almost only the fissile uneven Pu isotopes are brunt. Consequently, the ^{239}Pu share in the Pu vector at EOL is lower than in the fast reactor even if some new ^{239}Pu is bred from the ^{238}U matrix. In the MSFR, all Pu isotopes have almost reached their asymptotic values due to the long operational span. Thus the vector does not change very much between the configuration after 60 and 80 years. The ^{239}Pu content is in both cases much lower than in the solid fueled systems. However, another important point form the point of view of security is that the TRUs are newer required to be separated during the operation of a MSFR.

2.4. Consommation Accrue de Plutonium dans les Réacteurs à Neutrons Rapides/Consommation Accrue des Dechets dans les Reacteurs RApides (CAPRA/CADRA) Project Results

The CAPRA core design has been developed in the beginning of the 90 s. The design is based on the Superhénix arrangement, since in a first attempt it had been the plan to adopt Superphénix for the burning of plutonium. Thus the power is given to 1500 MWe or 3600 MWth [40]. The design problems are described by Broeders *et al.* [41]: "At present the Pu content is limited to 45% to remain compatible with the existing manufacturing processes. These high plutonium contents require high dilutions of the fuel in the core. These are obtained by a three-level dilution strategy: small diameter fuel pins with hollow fuel pellets, heterogeneous fuel assemblies containing about 30 'fuel-free' pins and a heterogeneous core layout with a number of 'fuel-free' sub-assemblies. A main problem for the core-layout is the high burnup reactivity, compared to existing LMFRs". The burnup for the study has been fixed to 200 GWd/tHM [8] or to 150 GWd/tHM [33]. The reactivity loss over the cycle of 285 d has been calculated to ~9000 pcm which corresponds to more than 27 $. In the study the Pu burning rat has been determined to 74.2 kg/TW_eh (≙ 30.9 kg/TW_{th}h) with a minor actinide production of 9.7 kg/TW_eh [8]. The results have been confirmed by Broeders *et al.*: "Accepting the above enrichment constraint [45%], an incineration rate of roughly 70 kg (Pu)/TWh(e) or 600 kg (Pu)/GWa(e) [≙ 28.5 kg/TW_{th}h] was obtained for a CAPRA-type reactor with MOX fuel" [41]. In a later study the plutonium burning in a Thorium based core has been calculated to 82 kg/TW_eh but with a production of a significant amount of U-233 [33].

During the studies at FZK some additional effects of the high Pu content have been discovered: "It turns out that for the CAPRA reactor with (U,Pu)-MOX fuel the upper limit of $Pu/(U + Pu) \approx 0.45$ works against quantitative recycling of burner Pu because of its quality decreasing with continued recycling" [42]. Consequently, this leads to the following conclusion: "CAPRA burner turns out to be a high-grade consumer of 'good and bad quality' Pu, leaving behind Pu which is deteriorated and proliferation resistant. In equilibrium, 58% of Pu from reprocessing of spent CAPRA fuel has to be ultimate excluded from recycling" [42]. Thus, the CAPRA 4/94 core is not sustainable. An unlimited reprocessing cannot be achieved since the Pu quality is rapidly degrading—this can be evaluated as good for the proliferation resistance, but not for the sustainable burning of plutonium. The degradation of the plutonium vector will lead to a configuration with which it will not be possible to achieve criticality in the follow up core.

CADRA Dedicated fast reactor with fertile free core with very low TRU mass in core (8 t, 88% Pu, 8% Am, 4% Cm) requires a high recycle mass flow. In this kind of reactors it is promised to achieve of a PU burning rate of 99 kg/TW_eh and a MA burning rate of 7 kg/TW_eh. [43] These two value sum to the theoretically possible value of 106 kg/TW_eh, or ~44 kg/TW_{th}h. The numbers for the specific TRU vector in this study are theoretical numbers without core design and feasibility of fuel production. The described core contains a very low Pu mass, but a very high Pu content and a cycle length of 480 d which will lead to a tremendous reactivity loss over cycle in a solid fuelled reactor. The problem of Pu degradation described above has to be expected for these systems, too.

Regarding dedicated cores, the experts of the FZK, now KIT, found the following burning rates to be reasonably achievable in a specially designed dedicated LMFR core for transmutation: "The following rates may be achieved: plutonium; 600 kg/GWea (≙ 28.5 $kg/TW_{th}h$), Np 237; 140 kg/GWea (≙ 6.7 $kg/TW_{th}h$), and Am 241; 80 kg/GWea (≙ 3.8 $kg/TW_{th}{\cdot}h$)" [41]. However, it has to be kept in mind that the results for CADRA as well as for CAPRA would only be achievable with a multiple recycling strategy and thus multiple reprocessing and fuel production which will be a growing challenge in time as well as in complexity with increasing number of steps.

Other innovative systems following the ideas discussed in CADRA using multiple recycling are under discussion and development. Some of them have been collected in the framework of INPRO of the IAEA [44] like the ABR [45] (advanced burner reactor) of the Global Nuclear Energy Partnership (GENEP) program of the United States or the accelerator driven system (ADS) [46] under development in several framework programs of the European community. The systems are discussed under the view of special fertile free fuels following the findings of CADRA like inert matrix fuel, rock-like oxide fuel, U-free metal fuel. Additionally, there are some commercial developments like PRISM [47] developed by GE Hitachi and RBWR [48,49] under development at Hitachi.

In general it has to be kept in mind for solid fueled reactors: "Fertile free fuel causes a high reactivity loss over the cycle due to the absence of a fertile material, since no breeding of new fissile material occurs... In fertile free fuel this excess reactivity has to be high to reach an acceptable long cycle time when no breeding takes place to produce fresh fissile material during the cycle like it is done in traditional fast reactor designs. Two choices are possible for the designer. On the one hand it would be possible to reduce the cycle time this results in reduced operational performance and prolonged out of operation time of the reactor... On the other hand it is possible to use a high excess reactivity which requires a strong control system with effective single control elements to compensate the excess reactivity at begin of cycle or a very high number of costly and complex control elements which have to be placed in the narrow space above the core" [14].

The results for the Pu reduction planned to be achieved in the CAPRA cores (~30 $kg/TW_{th}h$) can be exceeded in a molten salt fast reactor applying a fertile free configuration with a burning rate of ~43 $kg/TW_{th}h$. This value is within the expected accuracy to be seen as identical with the promised values of CADRA for the fertile free system. The exact theoretical number depends on the exact fuel configuration since the energy per fission is slightly isotope dependent.

The often discussed possibility of the Pu burning in MOX fuels in LWRs is real solution. The use of one cycle MOX in a LWR does not lead to a long term solution of the plutonium problem. Multiple recycling could achieve a balanced state, but only after several decades [50].

3. CONCLUSIONS

Pu burning will be an essential point to increase the sustainability of LWR operation for the future. At least as long as the closed fuel cycle option with massive use of fast reactors is not established as a standard. The Pu burning could avoid the accumulation of Pu in the final disposal. With this strategy the sometimes postulated Pu mine will be effectively avoided, even after the self-protection standard of the nuclear fuel has decayed.

MSRs offer very special features which make these kinds of reactors very much suitable for efficient burning of plutonium. One major key is the possibility of different operational modes based on fertile free fuel compositions. On the one hand this is possible due to the excellent inherent safety effects of MSFRs which are given due to the unity of coolant and fuel. On the other hand due to the high flexibility in the use of different fissile materials or qualities of the fissile material like it has been demonstrated in the MSRE by the smooth transition for ^{235}U fuel to ^{233}U fuel. The general operation management can be based on the strategy that the plutonium and other fissile materials which are put into the reactor core come never out anymore. In addition, this strategy is excellent from the point of view of nuclear security.

The comparison of the neutronic results with current state reactors like LWR with MOX and SFR shows the advantage of the MSFR for the Pu reduction. The main reason is the possibility of a fertile free core configuration. The detailed comparison with one of the first promising coordinated scientific project on the burning of TRU isotopes and plutonium demonstrates that the simulated MSFR could outperform the results gained in the CAPRA project from neutronic point of view. The CAPRA results of Pu burning rates of 28.5 kg/TW_{th}h to 30.9 kg/TW_{th}h can be exceeded by the fertile free configuration which enables to achieve the theoretical limit value of ~43 kg/TW_{th}h. This is the limit given by the average fission energy content of the different Pu isotopes. In the CADRA project it had been projected to achieve this limit, too by the application of a fertile free fuel configuration. However, it is from economic as well as from safety point of view almost impossible to operate a solid fueled critical reactor with a fertile free fuel configuration.

However, there is another nice effect due to the flexibility of MSFRs. If breeding is welcome in the fuel cycle strategy, it would be possible in MSFR, too. The big advantage of the MSFR in this case would be the elimination of the multi recycling and the solid fuel production. Additionally, no Pu, neither the inserted one, nor the bred one will come out of the reactor. Remember the statement: When the plutonium or other fissile material is once put into a molten salt reactor system it is not necessary to separate the material in any of the steps of the burning operation anymore.

4. OUTLOOK

Ideally, the idea of efficient TRU burning in a MSFR should end up in "our dream", an international endeavor carried by all countries with national strategies which do not envisage an extensive use of the fast reactor technology for energy production. This would be a way to stop the worldwide TRU accumulation problem in the future and thus to improve sustainability of LWR operation. The international endeavor for burning TRUs would be in our view the preferred solution compared to the transfer of the TRUs inside the burnt fuel into a final disposal (direct disposal). The burning of TRU is the only way to exclude misuse and theft of plutonium on the long term. We are not alone with this view; see the statement of Oi in the IAEA bulletin in 1998. The statement has already been given as opener. It is repeated here, because of its future relevance: "Within 200 years, the protection afforded by intense radioactivity (spent fuel standard) will disappear as the result of the decay of most radioactive nuclides. If the spent fuel is buried in a geological formation, it might be regarded as a potential 'plutonium mine' meaning that at some later point in time the buried plutonium could be mined and extracted" [2]. The application of a strategy to burn the existing plutonium efficiently would increase the sustainability of LWR operation by:

- Reducing the waste amount since Pu is the major volumetric challenge of P&T.

- Elimination the effects of the plutonium which has been produced during LWR operation, we should burn what we have built or produced. We brought the plutonium into the world and we should be able to clear it away efficiently.

- Elimination of the risk of misuse and theft of Pu from the burnt fuel after the self-protection has disappeared. There should be a better solution than to dig it only.

- Using the large energy amount latent in the plutonium to create electricity while we clear it away.

In an ideal case this objectives should be achieved:

- Without extensive multi-recycling, let's limit the reprocessing to one cycle, avoiding the solid fuel production at all and significantly reduce the transports which are one of the most vulnerable points which can develop to a condensation point of public protests like it appeared in Germany.
- Without separation of the plutonium in the several stages of the multi-recycling, what is once put into the reactor will stay there until it is burnt.
- Without installation of a classical fast reactor park for electricity production, thus without the multi-recycling and the often cited entering into so-call plutonium economy which could lead to public protests, too.

In our view, the general aim of all kinds of nuclear reactor operation should be to close the lifecycle of the plutonium from the cradle to the grave.

AUTHOR CONTRIBUTIONS

Study design, calculations and writing: Bruno Merk; calculations and writing: Dzianis Litskevich.

CONFLICTS OF INTEREST

The authors declare no conflict of interest.

REFERENCES

1. Arkhipov, V. Future nuclear energy systems: Generating electricity, burning wastes. IAEA Bulletin, 39/2/1997, 1997. Available online: https://www.iaea.org/sites/default/files/publications/magazines/bulletin/bull39-2/39204783033.pdf(accessed on 27 February 2015).
2. Oi, N. Plutonium Challenges—Changing Dimensions of Global Cooperation, IAEA Bulletin 40-1, Comparing Energy Options, 1998. Available online: http://www.iaea.org/Publications/Magazines/Bulletin/Bull401/article3.html (accessed on 26 March 2015).

3. Chow, B.G. Limiting the Accumulation of Weapon-Usable Plutonium, RAND Testimony, CT-177, 1994. Available online: http://www.rand.org/content/dam/rand/pubs/.../CT117.pdf (accessed on 23 January 2015).

4. Renn, O. (Hrsg.). Partitionierung und Transmutation. Forschung-Entwicklung-Gesellschaftliche Implikationen (acatech STUDIE), München: Herbert Utz Verlag 2014. Available online: http:// www.acatech.de/fileadmin/user_upload/Baumstruktur_nach_Website/Acatech/root/de/Projekte/Laufende_Projekte/Transmutation/PuT_Studie_komplett_2013-10-22.pdf (accessed on 27 November 2014). (In German).

5. Weitze, M.-D. Partitionierung und Transmutation nuklearer Abfälle. Chancen und Risiken in Forschung und Anwendung (acatech POSITION), München: Herbert Utz Verlag 2014. Available online: http://www.acatech.de/de/publikationen/stellungnahmen/acatech/detail/artikel/partitionierung-und-transmutation-nuklearer-abfaelle-chancen-und-risiken-in-forschung-und-anwend.html (accessed on 6 March 2015). (In German).

6. Merz, E.R. Introduction to NATO advanced research workshop. In *Advanced Nuclear Systems Consuming Excess Plutonium*, Proceedings of the Advanced Research Workshop on Advanced Nuclear Systems Consuming Excess Plutonium, Moscow, Russia, 13–16 October 1996; Merz, E.R., Walter, C.R., Eds.; Springer: Amsterdam, The Netherlands, 1997; Volume 15. [Google Scholar]

7. Puthiyavinayagam, P. Progress in Fast Reactor Programme of India: April 2014–March 2015. In Proceedings of the 48th Annual Meeting of TWGFR, IAEA IPPE, Obninsk, Russia, 25–29 May 2015.

8. Languille, A.; Gamier, J.C.; Lo Pinto, P.; Na, B.C.; Verrier, D.; Deplaix, J.; *et al*. CAPRA core studies. The oxide reference option. In Proceedings of the International Conference on Evaluation of Emerging Nuclear Fuel Cycle Systems, Versailles, France, 11–14 September 1995; pp. 874–881.

9. Gat, U.; Engel, J.R.; Dodds, H.L. Molten Salt Reactors for Burning Dismantled Weapons Fuel, Technical Note. *Nucl. Technol.* **1992**, *100*, 390–394. [Google Scholar]

10. MacPherson, H.G. The Molten Salt Reactor Adventure. *Nucl. Sci. Eng.* **1985**, *90*, 374–380. [Google Scholar]

11. Engel, J.R.; Grimes, W.R.; Rhoades, W.A.; Dearing, J.F. Molten-Salt Reactors for Efficient Nuclear Fuel Utilization without Plutonium Separation,

ORNL/TM-6413. August 1978. Available online: http://web.ornl. gov/info/reports/1978/3445603227167.pdf (accessed on 15 July 2015). [Google Scholar]

12. Engel, J.R.; Grimes, W.R.; Rhoades, W.A.; Dearing, J.F. Molten-Salt Reactors for Efficient Nuclear Fuel Utilization without Plutonium Separation. *Nucl. Technol.* **1979**, *46*, 30–43. [Google Scholar]

13. Merk, B.; Rohde, U.; Scholl, S. The Molten Salt Fast Reactor as Transmutation System. In Proceedings of the International Conference on Fast Reactors and Related Fuel Cycles: Safe Technologies and Sustainable Scenarios (FR13), Paris, France, 4–7 March 2013.

14. Merk, B.; Rohde, U.; Glivici-Cotruta, V.; Litskevich, D.; Scholl, S. On the Molten Salt Fast Reactor for Applying an Idealized Transmutation Scenario for the Nuclear Phase Out. *PLoS ONE* **2014**, *9*. [Google Scholar] [CrossRef] [PubMed]

15. Alekseev, P.N. Utilization of excess Plutonium in molten salt reactors. In *Advanced Nuclear Systems Consuming Excess Plutonium*, Proceedings of the Advanced Research Workshop on Advanced Nuclear Systems Consuming Excess Plutonium, Moscow, Russia, 13–16 October 1996; Merz, E.R., Walter, C.R., Eds.; Springer: Amsterdam, The Netherlands, 1997. [Google Scholar]

16. Kim, T.K.; Taiwo, T.A.; Wigeland, R.A.; Dixon, B.W.; Gehin, J.C.; Todosow, M. Mass Flow Data Comparison for Comprehensive Fuel Cycle Options. In Proceedings of the Actinide and Fission Product Partitioning and Transmutation Thirteenth Information Exchange Meeting, Seoul, Korea, 23–26 September 2014.

17. Renault, C.; Delpech, M. *MOST Final Report*; EURATOM Contract Number FIKI-CT-2001-20183; European Commission: Brussels, Belgium, 2005. [Google Scholar]

18. Mathieu, L.; Heuer, D.; Billebaud, A.; Brissot, R.; Garzenne, C.; Le Brun, C.; Lecarpentier, D.; Liatard, E.; Loiseaux, J.-M.; Meplan, O.; *et al.* Proposal for a Simplified Thorium Molten Salt Reactor. In Proceedings of GLOBAL 2005, Tsukuba, Japan, 9–13 October 2005; pp. 9–13.

19. EVOL—Evaluation and Viability of Liquid Fuel Fast Reactor System. Available online: http://cordis.europa.eu/search/index.cfm?fuseaction=proj.document&PJ_LANG=EN&PJ_RCN=11669355&pid=5 (accessed on 18 October 2012).

20. Renault, C.; Guérard, C. The Molten Salt Reactor (MSR). GIF System Development Progress Status. In Proceedings of the 4th INPRO-GIF Interface Meeting, Vienna, Austria, 1–3 March 2010.

21. Merle-Lucotte, E.; Heuer, D.; Allibert, M.; Brovchenko, M.; Capellan, N.; Ghetta, V. Launching the thorium cycle with molten salt reactor. In Proceedings of the International Congress on Advances in Nuclear Power Plants (ICAPP 2011), Nice, France, 2–5 May 2011.

22. Ingatiev, V.; Feynberg, O.; Merzlyakov, A.; Surenkov, A.; Zagnitko, A.; Afonichkin, V.; Bovet, A.; Khokhlov, V.; Subbotin, V.; Fazilov, R.; *et al.* Progress in Development of MOSART Concept with Th Support. In Proceedings of the International Congress on Advances in Nuclear Power Plants (ICAPP'12), Chicago, IL, USA, 24–28 June 2012.

23. Ignatiev, V.; Feynberg, O.; Gnidoi, I.; Merzlyakov, A.; Smirnov, V.; Surenkov, A.; Tretiakov, I.; Zakirov, R. Progress in Development of Li, Be, Na/F Molten Salt Actinide Recycler & Transmuter Concept. In Proceedings of the International Congress on Advances in Nuclear Power Plants (ICAPP) 2007, Nice, France, 13–18 May 2007.

24. Ignatiev, V.; Feynberg, O.; Gnidoi, I.; Merzlyakov, A.; Surenkov, A.; Uglov, V.; Zagnitko, A.; Subbotin, V.; Sannikov, I.; Toropov, A.; *et al.* Molten salt actinide recycler and transforming system without and with Th-U support: Fuel cycle flexibility and key material properties. *Ann. Nucl. Energy* **2014**, *64*, 408–420.

25. Ignatiev, V.; Feynberg, O.; Gnidoi, I.; Konakov, S.; Kormilitsyn, M.; Merzliakov, A.; Surenkov, A.; Uglov, V.; Zagnitko, A. MARS: Story on Molten Salt Actinide Recycler and Transmuter Development by Rosatom in Co-operation with Euratom. In Proceedings of the Actinide and Fission Product Partitioning and Transmutation Thirteenth Information Exchange Meeting, Seoul, Korea, 23–26 September 2014.

26. The raw data of the calculations and the used PYTHON script are stored at the HZDR library. Available online: https://www.hzdr.de/db/ !FzrTools. Archiving.ArchiveInfos?pNid=223&pId=1548 (accessed on 15 July 2015).

27. Brovchenko, M.; Merle-Lucotte, E.; Rouch, H.; Alcaro, F.; Aufiero, M.; Cammi, A.; Dulla, S.; Feynberg, O.; Firma, L.; Geoffroy, O.; *et al.* Evaluation and Viability of Liquid Fuel Fast Reactor System EVOL, Optimization of the pre-conceptual design of the MSFR, DELIVERABLE D 2.2. Available online: http://www.janleenkloosterman.nl/reports/evol_d22_201309.pdf (accessed on 30 October 2015).

28. Wemple, C.A.; Simeonow, T.; Rhodes, J.D., III. *HELIOS-2 Methods Version 2.1*; SSP-11/452 Rev 1, Studsvik Scandpower, Inc.: Idaho Falls, ID, USA; 16; December; 2011.

29. Merk, B.; Litskevich, D. Transmutation of all German Transuranium under Nuclear Phase out conditions—Is this Feasible from Neutronic Point of View? *PLoS One*. under review.

30. IAEA Fast Reactor Database—2006 Update. IAEA TECDOC 1531. Available online: http://www-pub.iaea.org/books/IAEABooks/7581/Fast-Reactor-Database-2006-Update (accessed on 27 October 2015).

31. Waltar, A.E.; Reynolds, A.B. *Fast Breeder Reactors*; Pergamon Press: New York, NY, USA, 1981.

32. Pay, A.; Francillon, E.; Steinmetz, B.; Barnes, D.; Meda, N. European Fast Reactor (EFR) Fuel Element Design. In Proceedings of the 10th International Conference on Structural Mechanics in Reactor Technology, Anaheim, CA, USA, 14–18 August 1989.

33. Salvatores, M.; Delpech, M. Advanced systems for Plutonium utilization, published in Advanced nuclear systems consuming excess plutonium. In *Advanced Nuclear Systems Consuming Excess Plutonium*, Proceedings of the Advanced Research Workshop on Advanced Nuclear Systems Consuming Excess Plutonium, Moscow, Russia, 13–16 October 1996; Merz, E.R., Walter, C.R., Eds.; Springer: Amsterdam, The Netherlands, 1997.

34. Kozlowski, T.; Downar, T.J. OECD/NEA AND U.S. NRC PWR MOX/UO2 CORE TRANSIENT BENCHMARK, OECD Nuclear Energy Agency Nuclear Science Committee, Final Specifications, Revision 2 December 2003. Available online:https://www.oecd-nea.org/science/wprs/MOX-UOX-transients/ (accessed on 27 October 2015).

35. Villarino, E.A.; Stammler, R.J.J.; Ferri, A.; Casal, J.J. HELIOS: Angularly dependent collision probabilities. *Nucl. Sci. Eng.* **1992**, *112*, 16–31.

36. Rachamin, R.; Wemple, C.; Fridman, E. Neutronic analysis of SFR core with HELIOS-2, Serpent, and DYN3D codes.*Ann. Nucl. Energy* **2013**, *55*, 194–204.

37. Brovchenko, M. Neutronic Benchmark of the Molten Salt Fast Reactor in the Frame of the EVOL and MARS Collaborative Projects. *Prog. Nucl. Energy*. under review.

38. Mikityuk, K. Core Safety (ESNII + WP6). In Proceedings of the European Nuclear Gen II, III, IV Days, Brussels, Belgium, 17–19 March 2015.

39. Merk, B.; Stanculescu, A.; Chellapandi, P.; Hill, R. Progress in reliability of fast reactor operation and new trends to increased inherent safety. *Appl. Energy* **2015**, *147*.

40. Rouault, J.; Garnier, J. C.; Languille, A.; Lo Pinto, P. The design of U-free large fast reactor cores. The CAPRA programme trends PHYSOR 96: Breakthrough of Nuclear Energy by Reactor Physics. In Proceedings of the International Conference on the Physics of Reactors, Mito, Japan, 16–20 September 1996.

41. Broeders, C.H.M.; Broeders, I.; Kessler, G.; Kiefhaber, E. Recent Neutron Physics Investigations on the Incineration of Plutonium and Other Transurania Isotopes. In *Advanced Nuclear Systems Consuming Excess Plutonium*, Proceedings of the Advanced Research Workshop on Advanced Nuclear Systems Consuming Excess Plutonium, Moscow, Russia, 13–16 October 1996; Merz, E.R., Walter, C.R., Eds.; Springer: Amsterdam, The Netherlands, 1997.

42. Wiese, H.W. Actinide transmutation properties of thermal and fast fission reactors including multiple recycling. *J. Alloys Compd.* **1998**, *271–273*, 522–529.

43. Vasile, A.; Vambenepe, G.; Lefèvre, J.C.; Hesketh, K.; Mashek, W.; De Raedt, C.; Haas, D. The CAPRA—CADRA programme, ICONE 8. In Proceedings of the 8th International Conference on Nuclear Engineering, Baltimore, MD, USA, 2–6 April 2000.

44. Status and Trends of Nuclear Technologies. Report of the International Project on Innovative Nuclear Reactors and Fuel Cycles (INPRO). Available online: http://www-pub.iaea.org/MTCD/publications/PDF/TE_1622_Web.pdf(accessed on 15 October 2015).

45. Global Nuclear Energy Partnership Technology Demonstration Program Dave Hill. Available online: http://www.mcs.anl.gov/events/ workshops/ nprcsafc/Presentations/NucPhysConf.pdf (accessed on 15 October 2015).

46. Knebel, J.; Eurotrans, I.P. A European Research Programme for the Transmutation of High Level Nuclear Waste in an Accelerator Driven System. Available online: https//www.oecd-nea.org/pt/docs/iem/ lasvegas04/11_Session_V/S5_01.pdf (accessed on 15 October 2015).

47. PRISM: Technology to Power and Secure our Energy Future. Available online: http://gehitachiprism.com/ (accessed on 15 October 2015).

48. Yamashita, J.; Kawamura, F.; Mochida, T. Next-generation Nuclear Reactor Systems for Future Energy. Available online: http://www.hitachi.com/rev/pdf/2004/r2004_03_105.pdf (accessed on 15 October 2015).

49. Takeda, R.; Aoyama, M.; Moriwaki, M.; Uchikawa, S.; Yokomizo, O.; Ochiai, K. General Features of Resource-Renewable BWR (RBWR) and Scenario of Long-term Energy Supply. In Proceedings of the International Conference on Evaluation of Emerging Nuclear Fuel Cycle Systems (GLOBAL95), Versailles, France, 11–14 September 1995.

50. Kessler, G.; Kim, J.I.; Gompper, K. Wohin mit dem deutschen Plutonium. *Int. Z. Kernenerg.* **1999**, *44*, 156–164. (In German).

Index